Thermoelectric Micro/Nano Generators 1

SCIENCES

Energy, Field Directors – Alain Dollet and Pascal Brault
Energy Recovery, Subject Head – Gustavo Ardila

Thermoelectric Micro/Nano Generators 1

Fundamental Physics, Materials and Measurements

Coordinated by
Hiroyuki Akinaga
Atsuko Kosuga
Takao Mori
Gustavo Ardila

WILEY

First published 2023 in Great Britain and the United States by ISTE Ltd and John Wiley & Sons, Inc.

Apart from any fair dealing for the purposes of research or private study, or criticism or review, as permitted under the Copyright, Designs and Patents Act 1988, this publication may only be reproduced, stored or transmitted, in any form or by any means, with the prior permission in writing of the publishers, or in the case of reprographic reproduction in accordance with the terms and licenses issued by the CLA. Enquiries concerning reproduction outside these terms should be sent to the publishers at the undermentioned address:

ISTE Ltd
27-37 St George's Road
London SW19 4EU
UK

www.iste.co.uk

John Wiley & Sons, Inc.
111 River Street
Hoboken, NJ 07030
USA

www.wiley.com

© ISTE Ltd 2023

The rights of Hiroyuki Akinaga, Atsuko Kosuga, Takao Mori and Gustavo Ardila to be identified as the authors of this work have been asserted by them in accordance with the Copyright, Designs and Patents Act 1988.

Any opinions, findings, and conclusions or recommendations expressed in this material are those of the author(s), contributor(s) or editor(s) and do not necessarily reflect the views of ISTE Group.

Library of Congress Control Number: 2023942071

British Library Cataloguing-in-Publication Data
A CIP record for this book is available from the British Library
ISBN 978-1-78945-144-3

ERC code:
PE8 Products and Processes Engineering
 PE8_6 Energy processes engineering

Contents

Preface ix
Hiroyuki AKINAGA, Atsuko KOSUGA and Takao MORI

Part 1. Introduction to Materials Development 1

Chapter 1. Strategies for Development of High Performance Thermoelectric Materials 3
Takao MORI, Atsuko KOSUGA and Hiroyuki AKINAGA

    1.1. Introduction 3
    1.2. Selectively lowering the thermal conductivity 5
        1.2.1. Utilizing nanostructuring and defects 5
        1.2.2. Utilizing crystal structure and bonding 7
    1.3. Enhancing the Seebeck coefficient/power factor 8
    1.4. Outlook for materials development 11
    1.5. References 12

Chapter 2. Computational and Data-Driven Development of Thermoelectric Materials. 17
Prashun GORAI and Michael TORIYAMA

    2.1. General theory 18
        2.1.1. Boltzmann transport theory 19
        2.1.2. Relaxation time approximation 20
        2.1.3. Thermoelectric properties 21
        2.1.4. Defect theory 24
    2.2. Applications 26
        2.2.1. Transport calculations 26
        2.2.2. Defect and doping calculations 40
        2.2.3. Thermoelectric material search with high-throughput computations and machine learning 45

2.3. Outlook 51
2.4. References 52

Part 2. Thermoelectric Materials 71

Chapter 3. Thermoelectric Copper and Silver Chalcogenides 73
Holger KLEINKE

3.1. Introduction 73
3.2. Binary copper and silver chalcogenides 75
3.3. Ternary and higher copper and silver chalcogenides 78
   3.3.1. Minerals based on copper and silver chalcogenides 78
   3.3.2. Tl-containing copper and silver chalcogenides 79
   3.3.3. Ba-containing copper and silver chalcogenides 79
3.4. Conclusion 84
3.5. Acknowledgments 85
3.6. References 85

Chapter 4. Sulfide Thermoelectrics: Materials and Modules 93
Michihiro OHTA, Priyanka JOOD and Kazuki IMASATO

4.1. Introduction 93
4.2. Materials 94
   4.2.1. Rare-earth sulfides 94
   4.2.2. Layered sulfides 97
   4.2.3. Pb–Bi–S-based systems 99
   4.2.4. Cu and Ag sulfide-based superionic conductors 101
   4.2.5. Tetrahedrites and colusites 103
   4.2.6. Chevrel-phase sulfides 105
   4.2.7. Chalcopyrite 106
4.3. Modules 107
   4.3.1. Colusites 107
   4.3.2. Cu and Ag sulfide-based superionic conductors 108
4.4. Summary and prospects 110
4.5. References 110

Chapter 5. A Concise Review of Strongly Correlated Oxides 125
Ichiro TERASAKI

5.1. Introduction to electron correlation 125
5.2. Electronic states of transition-metal oxides 129
5.3. 3D transition-metal oxides 130
   5.3.1. Co oxides 131
   5.3.2. Cu oxides 135
   5.3.3. Other 3D transition-metal oxides 136

5.4. 4D transition-metal oxides	136
5.4.1. Rh oxides	137
5.4.2. Ru oxides	139
5.5. Concluding remarks	139
5.6. References	140

Chapter 6. Nanocarbon Materials as Thermoelectric Generators — 149
Tsuyohiko FUJIGAYA and Yoshiyuki NONOGUCHI

6.1. Introduction	149
6.2. Carbon nanotubes	150
6.3. Transport to materials studies	151
6.4. Chemical doping	156
6.5. Thermoelectric generators using CNT	162
6.6. TEG based on CNT sheet	163
6.7. TEG fabrication based on CNT-based ink	169
6.8. CNT yarn and their fabric	172
6.9. Conclusion	175
6.10. References	176

Part 3. Metrology of Thermal Properties — 181

Chapter 7. Precise Measurement of the Absolute Seebeck Coefficient from the Thomson Effect — 183
Yasutaka AMAGAI

7.1. Introduction	183
7.2. Absolute scale of thermoelectricity	185
7.3. Measurement methods of the Thomson effect	189
7.3.1. Conventional method	190
7.3.2. New measurement methods: AC–DC method	192
7.4. Summary and outlook	195
7.5. References	196

Chapter 8. Thermal Diffusivity Measurement of Thin Films by Ultrafast Laser Flash Method — 201
Tetsuya BABA, Takahiro BABA and Takao MORI

8.1. Introduction	201
8.2. Laser flash method and ultrafast laser flash method	203
8.2.1. Laser flash method	203
8.2.2. Ultrafast laser flash method	205
8.3. Basic equation for data analysis	209
8.3.1. Response function method	209
8.3.2. Uniform single layer	212

8.3.3. Quadruple matrix 212
8.3.4. Thin film/substrate model 213
8.3.5. Temperature response after periodic pulse heating 216
8.4. Analysis of observed temperature response 222
8.4.1. Picosecond pulsed light heating 222
8.4.2. Nanosecond pulsed light heating 224
8.5. Metrological standard and traceability for measurements of thin film
thermophysical properties 224
8.6. Application of measurement from industrial to basic physics 225
8.7. References 226

List of Authors 233

Index 235

Summary of Volume 2 239

Preface

Hiroyuki AKINAGA[1], Atsuko KOSUGA[2] and Takao MORI[3,4]
[1] National Institute of Advanced Industrial Science and Technology (AIST),
Device Technology Research Institute, Japan
[2] Osaka Metropolitan University, Department of Physical Science,
Graduate School of Science, Japan
[3] National Institute for Materials Science (NIMS), WPI-MANA, Japan
[4] Graduate School of Pure and Applied Sciences, University of Tsukuba, Japan

Utilizing the Seebeck effect, thermoelectric materials can convert heat into electricity with a solid state device, i.e. a thermoelectric generator. This has the potential to contribute largely to society, considering around half of all primary energy consumed, i.e. fossil fuels, are lost in the form of waste heat, and the viable application of thermoelectric generators can lead to substantial energy saving. Furthermore, the trillions of sensors necessary for the Internet of Things (IoT) need a dynamic, maintenance-free power source of which thermoelectric generators are a promising candidate. The latter application is attractive because the motivation is not to supply cheap electricity, but high value electricity, which is suited for thermoelectrics, and also because thermoelectric conversion is viable with small size devices. With this background in mind, this two volume edition of *Thermoelectric Micro/Nano Generators* has been produced to serve as an important, comprehensive set of books, encompassing the fundamental principles, state-of-the-art advancements and outlooks for this topic.

Thermoelectric Micro/Nano Generators 1,
coordinated by Hiroyuki AKINAGA, Atsuko KOSUGA,
Takao MORI and Gustavo ARDILA.
© ISTE Ltd 2023.

In Volume 1, we mainly deal with fundamental physics, materials and measurements. After an introduction on the Strategies for Development of High Performance Thermoelectric Materials (Chapter 1) by several of the editors, a comprehensive chapter on Computational and Data-Driven Development of Thermoelectric Materials (Chapter 2) is given by Prashun Gorai and Michael Toriyama. They present in detail the general theory of carrier and phonon transport, also including defect formation and doping, which are powerful tools for thermoelectric enhancement. Recent powerful advancements in the computation field have been made regarding materials informatics and machine learning. The authors give detailed advice on how to use machine learning to accelerate the search for promising thermoelectric materials.

The next four chapters cover the detailed development and outlook of several promising thermoelecric material systems. Notable traditional high performance thermoelectric materials systems include Bi_2Te_3 and PbTe, which have unattractive features such as the extreme rareness of Te and high toxicity of Pb. Four more sustainable material systems are presented. Holger Kleinke covers Thermoelectric Copper and Silver Chalcogenides (Chapter 3). Cu_2S, Cu_2Se and their silver counterparts are abundant materials which have been reported to exhibit an extremely high number of merits. This chapter covers their basic features and recent advancements, also regarding their stability, which has been an issue, together with developments of several other selected chalcogenides. Michihiro Ohta, Priyanka Jood and Kazuki Imasato have provided a comprehensive chapter on Sulfide Thermoelectrics: Materials and Modules (Chapter 4). The versatility and high performance of a wide variety of sulfide materials, from layered sulfides like TiS_2-based compounds and misfit compounds, high temperature rare earth sulfides, Chevrel-phase sulfides and mineral-based sulfides, such as tetrahedrite, colusite, chalcopyrite, etc., and module fabrication and performance of some of the compounds, are covered. Ichiro Terasaki has provided A Concise Review of Strongly Correlated Oxides (Chapter 5). Oxides are some of the most inexpensive, stable and abundant compounds. Terasaki particularly focuses on the physics principles; namely the strong correlation and manifestation of the spin and orbital degrees of freedom in the thermoelectric properties, which have resulted in notable high performances. He systematically covers several intriguing systems in the 3D transition-metal oxides and 4D transition-metal oxides, while giving the wide context and meaning in terms of physical principles. Tsuyohiko Fujigaya and Yoshiyuki Nonoguchi cover the relatively recent emergence of Nanocarbon Materials as Thermoelectric Generators (Chapter 6). They particularly focus on the fundamentals and development of thermoelectric carbon nanotubes (CNTs). The production capability and quality control of CNT materials has advanced significantly to make it a candidate for one of the potentially most abundant and sustainable material families. The authors review in detail the thermoelectric properties of CNTs and control via doping, and also comprehensively cover the utilization of CNTs in various format thermoelectric generators, including details of processing such as via ink, etc.

The final two chapters deal with the metrology of thermal properties. Precise evaluation of the thermoelectric properties is critical for the effective development of materials and devices. Yasutaka Amagai covers the Precise Measurement of the Absolute Seebeck Coefficient from the Thomson Effect (Chapter 7). He includes an instructive and detailed introduction to the measurement principles of the Seebeck coefficent, and elucidates new methods regarding precision measurement of the absolute Seebeck coefficient, which is also known as the absolute scale of thermoelectricity. Tetsuya Baba, Takahiro Baba and Takao Mori have provided a chapter on Thermal Diffusivity Measurement of Thin Films by Ultrafast Laser Flash Method (Chapter 8). The accurate evaluation of thermal diffusivity, and thereby thermal conductivity of thin films, is a difficult topic, which is important not just for thermoelectrics, but for a variety of fields. The principles of the measurement of bulk materials is given, and advances in the ultrafast laser flash method utilizing thermoreflectance for thin film measurements are given in detail, including equations and derivations for accurate data analysis.

This first volume of *Thermoelectric Micro/Nano Generators* comprehensively covers the fundamentals and state-of-the-art advances in thermoelectric property enhancement, computational methodolgy, various promising material systems, and metrology, which should be useful to a wide range of readers, from people who have some curiosity regarding this topic and beginners in the field, up to experts desiring guidelines and new clues for advanced development.

July 2023

PART 1

Introduction to Materials Development

1
Strategies for Development of High Performance Thermoelectric Materials

Takao MORI[1,2], Atsuko KOSUGA[3] and Hiroyuki AKINAGA[4]

[1] *National Institute for Materials Science (NIMS), WPI-MANA, Japan*
[2] *Graduate School of Pure and Applied Sciences, University of Tsukuba, Japan*
[3] *Osaka Metropolitan University, Department of Physical Science, Graduate School of Science, Japan*
[4] *National Institute of Advanced Industrial Science and Technology (AIST), Device Technology Research Institute, Japan*

1.1. Introduction

Thermoelectric materials and modules/devices represent the solid-state conversion of waste heat to electricity (Rowe et al. 1995). Because of recent carbon neutral goals and the necessity of realizing the IoT (Internet of Things), thermoelectric generators will hopefully contribute to energy saving via waste heat power generation (Bell 2008; Hendricks et al. 2022), and also as dynamic power sources for innumerable sensors and devices (Mori and Priya 2018; Petsagkourakis et al. 2018; Yan et al. 2018; Akinaga 2020). To achieve these goals, it is vital to accelerate the development of both materials and modules, and applications. In this chapter, regarding the former, the materials development, we give an overview on basic strategies for the development of high performance thermoelectric materials.

Thermoelectric Micro/Nano Generators 1,
coordinated by Hiroyuki AKINAGA, Atsuko KOSUGA,
Takao MORI and Gustavo ARDILA.
© ISTE Ltd 2023.

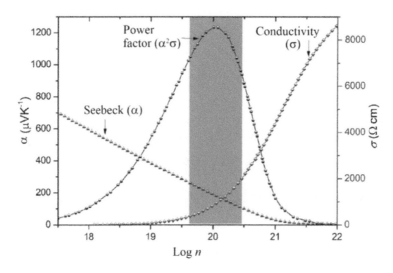

Figure 1.1. *Schematic of general trend of carrier concentration dependence of Seebeck coefficient, conductivity, power factor (Snyder and Toberer 2008)*

The performance of TE materials is gauged by the so-called figure of merit, $ZT = \alpha^2 \sigma T/\kappa$, where α is the Seebeck coefficient, σ is electrical conductivity, κ is thermal conductivity, and T is temperature. The numerator $\alpha^2 \sigma$ is called the power factor and is a measure of the maximum output power that can be obtained in a fixed applied temperature difference condition. The larger ZT is, the closer the maximum conversion efficiency η_{max} of the thermoelectric material approaches the ideal Carnot efficiency $\eta_{Carnot} = (T_h - T_c)/T_h$, where T_h and T_c are the hot side and cold side temperatures, respectively, following this dependency on ZT:

$$\eta_{max} = \eta_{Carnot}[(1 + ZT_m)^{1/2} - 1] / [(1 + ZT_m)^{1/2} + T_c/T_h], \qquad [1.1]$$

where T_m is the average temperature. Therefore, it is imperative to achieve as high a ZT as possible to obtain high performance thermoelectric performance. The difficulty is that the physical properties which compose ZT have interrelating tradeoffs. Namely for high ZT, it is ideal to have a large Seebeck coefficient α, like an insulator, but also a large σ, like a metal, namely with a high electrical conductivity, while simultaneously possessing low thermal conductivity κ. In conventional optimization, the carrier concentration of a material is tuned by doping. For example, to try to obtain an optimum balance between α and σ, as illustrated in Figure 1.1 (Snyder and Toberer 2008). This can lead to an increase in the power factor and ZT

in the realms of conventional optimization. In general, it is considered that the heavily doped semiconductor regime (e.g. carrier concentration in a wide range around 10^{20} cm^{-3}) is a good target for high performance thermoelectric materials.

However, power generation applications have not been widely realised yet, and it is strongly desirable to attain higher ZT and higher conversion efficiency. This dictates the need to develop enhancement principles that can overcome the tradeoffs, and in this chapter we will briefly review several of them.

1.2. Selectively lowering the thermal conductivity

1.2.1. *Utilizing nanostructuring and defects*

The thermal conductivity is composed of two components; thermal conductivity carried by electrical carriers κ_e and the lattice, i.e. phonons, κ_l, namely $\kappa = \kappa_e + \kappa_l$. From the Wiedemann-Franz relationship, κ_e is proportional to σ which is necessary for a high ZT, and therefore, the vital point is to reduce κ_l in a way that does not kill σ. Liu et al. (2013) demonstrated that the lattice thermal conductivity, κ_l, can be approximated from Callaway's frequency-dependent relaxation time approximation and effective medium approximation:

$$\kappa_l^{-1} = \kappa_0^{-1} + 2R_K d^{-1}, \qquad [1.2]$$

where κ_0 represents the bulk thermal conductivity of the material, R_K the interfacial thermal resistance, i.e. Kapitza resistance, and d the grain diameter. Therefore, the grain diameter can be reduced via nanomicrostructuring to obtain a reduced κ_l. One point to be careful about is that the size should preferably be controlled, i.e. made not too small, so that electrical carriers are not overly scattered. Namely, utilizing the difference in magnitude in the mean-free-paths of electrical carriers and phonons, so that the phonons are selectively scattered by the grain boundaries. It should also be stressed that the phonon mean-free-path can have a large variation depending on the particular material (Liu et al. 2013).

There are many ways to achieve this nanomicrostructuring, from top-down methods such as ball-milling, super-plastic deformation (SPD), high pressure torsion (HPT), etc., to bottom-up methods such as the wet fabrication of nanomicroparticles, creating precipitations, nanopores, etc. in the material. Many useful reviews have been written on the details of this topic (Sootsman et al. 2009; Mori 2017; Liu et al. 2018; Mao et al. 2018).

One point which should be stressed is that the understanding of how these various defects scatter phonons in different ways has greatly advanced (Liu et al. 2018). Namely, the phonon scattering rate τ^{-1}, which is the inverse of phonon relaxation time τ, has a different phonon frequency ω depending on the different types of defects. For point defects, namely, alloyed/doped atoms, vacancies, interstitials:

$$\tau_{PD}^{-1} \propto \omega^4, \qquad [1.3]$$

which shows that these defects are particular effective to scatter high frequency phonons. For line defects, namely, dislocation strain fields and dislocation cores:

$$\tau_{DS}^{-1} \propto \omega^1, \; \tau_{DC}^{-1} \propto \omega^3, \qquad [1.4]$$

respectively, indicating these defects scatter mid-frequency phonons well. Two dimensional defects, namely, grain boundaries:

$$\tau_{GB}^{-1} \propto \omega^0, \qquad [1.5]$$

scatter low frequency phonons. Volume defects, namely, nanopores and nanoprecipitates:

$$\tau_{VD}^{-1} \propto \omega^0 + \omega^4, \qquad [1.6]$$

scatter phonons over a wide range. Phonon scattering is relatively well understood, so the key for enhancement of each material is how to introduce appropriate defects/nanostructuring to effectively selectively lower the thermal conductivity.

One recent interesting effect discovered was that a very small amount of interstitial doping into Mg_3Sb_2-type material largely reduced the phonon group velocity, leading to radically low thermal conductivity and high ZT for this material (Liu et al. 2021a). An initial module demonstrated conversion efficiency comparable to the best Bi_2Te_3-type modules, which have been champions for over half a century (Liu et al. 2021a, 2022), underlining the effectiveness of this type of approach to achieve high performance.

1.2.2. Utilizing crystal structure and bonding

As reviewed in the previous section, utilizing nanostructuring and defects are effective extrinsic methods to lower the lattice thermal conductivity, κ_l. In regard to intrinsic effects, focusing on particular crystal structures and bonding features can provide ways to find inherent low thermal conductivity materials. Slack (in Seitz, F. et al. 1979) gave early guidelines, pointing out that crystal complexity can lead to low thermal conductivity. Namely assuming contribution of only acoustic phonons, κ_l can be approximated in the following formula,

$$\kappa_l \sim B M \delta (\theta_D)^3 n^{-2/3} T^{-1} \gamma^{-2}, \qquad [1.7]$$

where B is a constant, M the mean atomic mass, δ a length parameter equalling (volume of the primitive cell)$^{1/3}$, θ_D the Debye temperature, n the number of atoms in the primitive cell, and γ the Gruneisen constant. Therefore, structurally complex materials with a large number of atoms in the primitive unit cell exhibit low κ_l.

Slack (1995), Nolas et al. (1996), Uher (2021) and others also focused on so-called cage compounds, skutterudites and clathrates where loosely bound atoms in the structural voids, i.e. guest atoms, were proposed to have rattling, namely anharmonic vibrations which effectively scattered acoustic phonons and led to low thermal conductivity. On the other hand, the electronic bands formed by the framework cage structures provided relatively high electrical conductivity, leading to the development of high ZT compounds amongst these group of compounds (Rogl et al. 2015; Uher 2021). In addition to rattling, lone pairs in crystal structures have also been demonstrated to lead to low thermal conductivity (Nielsen et al. 2013; Du et al. 2016). Boron cluster compounds have also embodied several low thermal conductivity mechanisms, such as crystal complexity, as pointed out by Slack (Seitz et al. 1979) and Cahill et al. (1989), boron dumbell pairs, and partial occupancy (Mori et al. 2007; Mori 2019). Parital occupancy was also recently utilized as a descriptor in in a materials genome approach to unearth ~0.4 W/m/K low thermal conductivity compounds in other systems (Liu et al. 2021b). A copper sulfide with ZT~0.7 was developed from these discoveries.

Recently, bonding effects on phonon scattering have been particularly systematically focused on. In addition to particular resonant bonding which has been proposed for PbTe (Lee et al. 2014), In_2Te_5 (Zhang et al. 2020), etc., bonding heterogeneity has been widely noted and utilized. Namely, this can be considered to be widely related to all mechanisms that have been presented, where bonding heterogeneity, namely a mixture in a compound of weak and strong bonding, leads to anharmonicity and low κ_l. The effect of bonding heterogeneity has been noted in many systems up until now (Pal et al. 2018; Dutta et al. 2020). As a recent example, 8 fold difference in lattice thermal conductivity has been observed in similar systems with single or mixed anion compounds. The local distortion in the crystal structure and bonding because of the mixed anions has been found to cause peak splitting in the phonon density of states (DOS), which increases the scattering phase space, leading to dramatically reduced low κ_l (Sato et al. 2021).

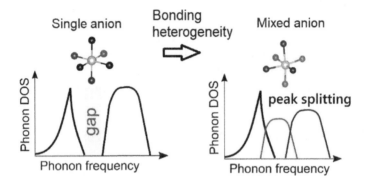

Figure 1.2. *Schematic of mixed anion effect on the phonon DOS (Sato et al. 2021). For a color version of this figure, see www.iste.co.uk/akinaga/thermoelectric1.zip*

1.3. Enhancing the Seebeck coefficient/power factor

From the Mott formula, the Seebeck coefficient α is proportional to the energy differential of the energy dependent conductivity σ at the Fermi level E_F (i.e. chemical potential), as shown in equation [1.8].

$$\alpha = \frac{\pi^2 k_B^2}{3e\sigma} T \left(\frac{\partial \sigma}{\partial E}\right)_{E_F}. \qquad [1.8]$$

In the single parabolic band approximation (SPB), α can be simply given as

$$\alpha = \frac{8\pi^2 k_B^2}{3eh^2} m^* T \left(\frac{\pi}{3n}\right)^{2/3}, \qquad [1.9]$$

where m^* and n are the effective mass of the charge carrier and the carrier concentration, respectively. A heavy effective mass m^* leads to a large Seebeck coefficient. This means that so-called *drag* effects, which will be descibed next, can enhance α. However, this is tempered by the fact that a large m^* generally leads to a low carrier mobility, and therefore, it is always paramount to keep an eye on maximizing the power factor $\alpha^2\sigma$.

There have been several Seebeck coefficient enhancement methods proposed which are based on modifying the band structure, i.e. band engineering.

The Seebeck coefficient α is correlated to the energy differential of DOS at the Fermi level, and an early proposal was made by Hicks and Dresselhaus (1993), in which they focused on the fact that low dimensionality can lead to sharp features in DOS, and proposed that the so-called confinement effect can enhance α. In another method, Heremans et al. (2008) proposed introducing resonance states via doping, thereby modify the curvature of DOS near the Fermi level. Wang et al. (2013) pointed out that the degeneracy of bands, i.e. band convergence, can also lead to enhanced Seebeck coefficients.

Band engineering from top down is not always easy to achieve. In contrast, as a method which is relatively easy to implement, it has been proposed to utilize magnetic ion doping, in which if the magnetic ion has strong coupling with the electrical carriers, the effective mass can be enhanced, and the overall power factor can be improved, as reviewed, for example, in (Mori 2017; Hendricks et al. 2022). This has been initially demonstrated in non-magnetic systems, such as $CuGaTe_2$ (Ahmed et al. 2017), Bi_2Te_3 (Vaney et al. 2019), etc. and called paramagnon drag (Zheng et al. 2019). On the overall topic of utilizing magnetism to enhance thermoelectrics, there has been a rich field of endeavors. Early, in some magnetic elemental materials like iron and nickel at low temperatures, enhancements, i.e. peaks, of the Seebeck coefficients were observed, and were attributed to magnon drag (Bailyn 1962). Magnon drag has also recently been shown to be effective at high temperatures in relatively high performance materials (Tsujii and Mori 2013; Ang et al. 2015; Matsuura et al. 2022). Terasaki (2013), pointed out that magnetic entropy was the source of enhanced performance in the cobaltites, originating from the spin high and low states of Co. This phenomenon has also been observed for some magnetic sulfides by Hebert et al. (2021). Recently, in weak itinerant ferromagnets, enhancement of the Seebeck coefficient at high temperatures such as 300 K and above, was indicated to occur from spin fluctuation (Tsujii et al. 2019). Namely, it can also intuitively and broadly be considered that entropy and spin fluctuation weigh down the carriers.

Phonon drag, in which strong interaction of the carriers with lattice, i.e. electron phonon interaction, was also observed to enhance α at an early time (Klemens 1954), more recently, dramatically lead to huge power factors in FeSb$_2$ at low temperatures (Bentien et al. 2006; Takahashi et al. 2016).

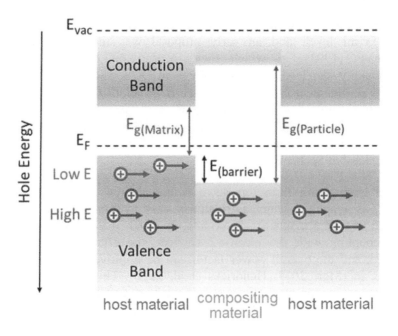

Figure 1.3. *Schematic of an energy filtering example of a p-type material to enhance the Seebeck coefficient (Pakdel et al. 2018). For a color version of this figure, see www.iste.co.uk/akinaga/thermoelectric1.zip*

It should be noted that one advantage in terms of possible application, is that for the recently discovered paramagnon drag and spin fluctuation enhancements, their effective temperature range can be higher than the conventional magnon drag and phonon drag (Hendricks et al. 2022).

In regard to other power factor enhancement principles, the so-called energy filtering phenomena has also been widely claimed. This occurs in materials composed of two or more components, where as shown in the schematic, the energetic barrier at the interfaces functions as a filter, scattering the low energy carriers and allowing the high energy carriers to selectively pass, which enhances the Seebeck coefficient (Shakouri et al. 1998; Heremans et al. 2004). Such enhancement has been claimed for many nanocompositing materials, and it is therefore a method that is relatively easy to at least attempt to implement.

For nanocomposites, a niche method showed that by introducing partially percolating metallic channels, this could lead to a large improvement of the electrical conductivity without killing the Seebeck coefficient in some cases, thereby enhancing the power factor (Mori and Hara 2016). This was demonstrated for relatively low σ, high α compounds, however, it still remains to be demonstrated whether high ZT can be achieved.

1.4. Outlook for materials development

As described above, there have been notable advancements in developing various strategies to enhance the performance of thermoelectric materials. For a long time, the champions in the low temperature region near room temperature and mid-high temperature region, have been Bi_2Te_3-type materials and PbTe-type materials. These materials are not optimum in that they have extremely tellerium as a main constituent, and for the latter, also lead which is a highly toxic element subject to restrictions. However, recently, earth abundant materials, such as sulfides, antimonides and carbon-based materials have been demonstrated to be able to exhibit high performance, expanding the possible target materials. Furthermore, the understanding of controlling the electronic and phononic transport and electronic structures has also advanced. These developments give hope that the materials development can lead to the long awaited wide-spread application of thermoelectric power generation.

1.5. References

Ahmed, F., Tsujii, N., Mori, T. (2017). Thermoelectric properties of $CuGa_{1-x}Mn_xTe_2$: Power factor enhancement by incorporation of magnetic ions. *J. Mater. Chem. A*, 5, 7545–7554.

Akinaga, H. (2020). Recent advances and future prospects in energy harvesting technologies. *Jpn. J. Appl. Phys.*, 59, 110201.

Ang, R., Khan, A.U., Tsujii, N., Takai, K., Nakamura, R., Mori, T. (2015). Thermoelectricity generation and electron–magnon scattering in a natural chalcopyrite mineral from a deep-sea hydrothermal vent. *Angew. Chem. Int. Ed.*, 54, 12909–12913.

Bailyn, M. (1962). Maximum variational principle for conduction problems in a magnetic field, and the theory of magnon drag. *Phys. Rev.*, 126, 2040–2054.

Bell, L.E. (2008). Cooling, heating, generating power, and recovering waste heat with thermoelectric systems. *Science*, 321, 1457–1461.

Bentien, A., Madsen, G.K.H., Johnsen, S., Iversen, B.B. (2006). Experimental and theoretical investigations of strongly correlated $FeSb_{2-x}Sn_x$. *Phys. Rev. B*, 74, 205105.

Cahill, D.G., Fischer, H.E., Watson, S.K., Pohl, R.O., Slack, G.A. (1989). Thermal properties of boron and borides. *Phys. Rev. B*, 40, 3254.

Du B., Chen, K., Yan, H., Reece, M.J. (2016). Efficacy of lone-pair electrons to engender ultralow thermal conductivity. *Scr. Mater.*, 111, 49–53.

Dutta, M., Samanta, M., Ghosh, T., Voneshen, D.J., Biswas, K. (2020). Evidence of highly anharmonic soft lattice vibrations in a Zintl rattler. *Angew. Chem. Int. Ed.*, 60, 4259–4265.

Hébert, S., Daou, R., Maignan, A., Das, S., Banerjee, A., Klein, Y., Bourgès, C., Tsujii, N., Mori, T. (2021). Thermoelectric materials taking advantage of spin entropy: Lessons from chalcogenides and oxides. *Sci. Technol. Adv. Mater.*, 22, 583–596.

Hendricks, T., Caillat, T., Mori, T. (2022). Keynote review of latest advances in thermoelectric generation materials, devices, and technologies 2022. *Energies*, 15, 7307.

Heremans, J.P., Thrush, C.M., Morelli, D.T. (2004). Thermopower enhancement in lead telluride nanostructures. *Phys. Rev. B*, 70, 115334.

Heremans, J.P., Jovovic, V., Toberer, E.S., Saramat, A., Kurosaki, K., Charoenphakdee, A., Yamanaka, S., Snyder, G.J. (2008). Enhancement of thermoelectric efficiency in PbTe by distortion of the electronic density of states. *Science*, 321, 554–557.

Hicks, L.D. and Dresselhaus, M.S. (1993). Effect of quantum-well structures on the thermoelectric figure of merit. *Phys. Rev. B*, 47, 12727.

Klemens, P.G. (1954). The contribution of phonons to the Thomson coefficient. *Aust. J. Phys.*, 7, 520.

Lee, S., Esfarjani, K., Luo, T., Zhou, J., Tian, Z., Chen, G. (2014). Resonant bonding leads to low lattice thermal conductivity. *Nat. Commun.*, 5, 3525.

Liu, W., Ren, Z., Chen, G. (2013). Nanostructured thermoelectric materials. In *Thermoelectric Nanomaterials*, Koumoto, K. and Mori, T. (eds). Springer, Berlin/Heidelberg.

Liu, Z., Mao, J., Liu, T.-H., Chen, G., Ren, Z. (2018). Nano-microstructural control of phonon engineering for thermoelectric energy harvesting. *MRS Bull.*, 43, 181–186.

Liu, Z., Sato, N., Gao, W., Yubuta, K., Kawamoto, N., Mitome, M., Kurashima, K., Owada, Y., Nagase, K., Lee, C.-H. et al. (2021a). Demonstration of ultrahigh thermoelectric efficiency of 7.3% in Mg_3Sb_2/MgAgSb module for low-temperature energy harvesting. *Joule*, 5, 1196–1208.

Liu, Z., Zhang, W., Gao, W., Mori, T. (2021b). A material catalogue with glass-like thermal conductivity mediated by crystallographic occupancy for thermoelectric application. *Energy Environ. Sci.*, 14, 3579–3587.

Liu, Z., Gao, W., Oshima, H., Nagase, K., Lee, C.-H., Mori, T. (2022). Maximizing the performance of N-type Mg_3Bi_2 based materials for room-temperature power generation and thermoelectric cooling. *Nat. Commun.*, 13, 1120.

Mao, J., Liu, Z., Zhou, J., Zhu, H., Zhang, Q., Chen, G., Ren, Z. (2018). Advances in thermoelectrics. *Adv. Phys.*, 67, 69–147.

Matsuura, H., Ogata, M., Mori, T., Bauer, E. (2021). Theory of huge thermoelectric effect based on a magnon drag mechanism: Application to thin-film Heusler alloy. *Phys. Rev. B*, 104, 214421.

Mori, T. (2017). Novel principles and nanostructuring methods for enhanced thermoelectrics. *Small*, 13, 1702013.

Mori, T. (2019). Thermoelectric and magnetic properties of rare earth borides: Boron cluster and layered compounds. *J. Solid State Chem.*, 275, 70–82.

Mori, T. and Hara, T. (2016). Hybrid effect to possibly overcome the trade-off between Seebeck coefficient and electrical conductivity. *Scr. Mater.*, 111, 44–48.

Mori, T. and Priya, S. (2018). Materials for energy harvesting: At the forefront of a new wave. *MRS Bull.*, 43, 176–180.

Mori, T., Martin, J., Nolas, G. (2007). Thermal conductivity of $YbB_{44}Si_2$. *J. Appl. Phys.*, 102, 073510.

Nielsen, M.D., Ozolins, V., Heremans, J.P. (2013). Lone pair electrons minimize lattice thermal conductivity. *Energ. Environ. Sci.*, 6, 570.

Nolas, G.S., Slack, G.A., Morelli, D.T., Tritt, T.M., Ehrlich, A.C. (1996). The effect of rare-earth filling on the lattice thermal conductivity of skutterudites. *J. Appl. Phys.*, 79, 4002.

Pakdel, A., Guo, Q.S., Nicolosi, V., Mori, T. (2018). Enhanced thermoelectric performance of $Bi-Sb-Te/Sb_2O_3$ nanocomposites by energy filtering effect. *J. Mater. Chem. A*, 6, 21341–21349.

Pal, K., He, J., Wolverton, C. (2018). Bonding hierarchy gives rise to high thermoelectric performance in layered Zintl compound $BaAu_2P_4$. *Chem. Mater.*, 30, 7760–7768.

Petsagkourakis, I., Tybrandt, K., Crispin, X., Ohkubo, I., Satoh, N., Mori, T. (2018). Thermoelectric materials and applications for energy harvesting power generation. *Sci. Technol. Adv. Mater.*, 19, 836–862.

Rogl, G., Grytsiv, A., Yubuta, K., Puchegger, S., Bauer, E., Raju, C., Mallik, R.C., Rogl, P. (2015). In-doped multifilled n-type skutterudites with ZT = 1.8. *Acta Mater.*, 95, 201.

Rowe, D.M. (1995). *CRC Handbook of Thermoelectrics*. CRC Press, Boca Raton.

Sato, N., Kuroda, N., Nakamura, S., Katsura, Y., Kanazawa, I., Kimura, K., Mori, T. (2021). Bonding heterogeneity in mixed-anion compounds realizes ultralow lattice thermal conductivity. *J. Mater. Chem. A*, 9, 22660–22669.

Seitz, F., Turnbull, D., Ehrenreich, H. (1979). *Semiconductors and Semimetals*, vol. 34. Academic Press, New York.

Shakouri, A., LaBounty, C., Abraham, P., Piprek, J., Bowers, J.E. (1998). Enhanced thermionic emission cooling in high barrier superlattice heterostructures. *MRS Proc.*, 545.

Slack, G.A. (1995). New materials and performance limits for thermoelectric cooling. In *CRC Handbook of Thermoelectrics*, Rowe, D.M. (ed.). CRC Press, Boca Raton.

Snyder, G.J. and Toberer, E.S. (2008). Complex thermoelectric materials. *Nat. Mater.*, 7, 105–114.

Sootsman, J., Young Chung, D., Kanatzidis, M.G. (2009). New and old concepts in thermoelectric materials. *Angew. Chem. Int. Ed.*, 48, 8616–8639.

Takahashi, H., Okazaki, R., Ishiwata, S., Taniguchi, H., Okutani, A., Hagiwara, M., Terasaki, I. (2016). Colossal Seebeck effect enhanced by quasi-ballistic phonons dragging massive electrons in $FeSb_2$. *Nat. Commun.*, 7, 12732.

Terasaki, I. (2013). Layered cobalt oxides: Correlated electrons for thermoelectrics. In *Thermoelectric Nanomaterials*, Koumoto, K. and Mori, T. (eds). Springer, Berlin/Heidelberg.

Tsujii, N. and Mori, T. (2013). High thermoelectric power factor in a carrier-doped magnetic semiconductor $CuFeS_2$. *Appl. Phys. Express*, 6, 043001.

Tsujii, N., Nishide, A., Hayakawa, J., Mori, T. (2019). Observation of enhanced thermopower due to spin-fluctuation in weak itinerant ferromagnet. *Sci. Adv.*, 5, eaat5935.

Uher, C. (2021). Thermoelectric properties of Skutterudites. In *Thermoelectric Skutterudites*. CRC Press, Boca Raton.

Vaney, J.-B., Yamini, S.A., Takaki, H., Kobayashi, K., Kobayashi, N., Mori, T. (2019). Magnetism-mediated thermoelectric performance of the Cr-doped bismuth telluride tetradymite. *Mater. Today Phys.*, 9, 100090.

Wang, H., Pei, Y., LaLonde, A.D., Snyder, G.J. (2013). Material design considerations based on thermoelectric quality factor. In *Thermoelectric Nanomaterials*, Koumoto, K. and Mori, T. (eds). Springer, Berlin/Heidelberg.

Yan, J., Liao, X., Yan, D., Chen, Y. (2018). Review of micro thermoelectric generator. *J. Microelectromech. Syst.*, 27, 1–18.

Zhang, W., Sato, N., Tobita, K., Kimura, K., Mori, T. (2020). Unusual lattice dynamics and anisotropic thermal conductivity in In_2Te_5 due to a layered structure and planar-coordinated Te-chains. *Chem. Mater.*, 32, 5335–5342.

Zheng, Y., Lu, T., Polash, M.M.H., Rasoulianboroujeni, M., Liu, N., Manley, M.E., Deng, Y., Sun, P.J., Chen, X.L., Hermann, R.P. (2019). Paramagnon drag in high thermoelectric figure of merit Li-doped MnTe. *Sci. Adv.*, 5, eaat9461.

2

Computational and Data-Driven Development of Thermoelectric Materials

Prashun GORAI[1] and Michael TORIYAMA[2]

[1]*Colorado School of Mines, Golden, CO, USA*
[2]*Northwestern University, Evanston, IL, USA*

The development of thermoelectric (TE) materials has witnessed decades of progress. Experiments, analytical modeling and computations have been combined with solid-state physics and chemistry to discover, design and optimize TE materials. In the 21st century, computational and data-driven approaches have played a pivotal role in the development of new TE materials and the optimization of existing ones. Additionally, computations have enabled a deeper fundamental understanding of charge carrier and thermal transport, which in turn has led to design rules for TE materials.

Edisonian approaches are currently the most common for new TE material discovery, but are likely to make only a small dent in exploring vast chemical spaces. Computations have complemented experimental efforts through modeling of transport properties, for example, with semi-classical Boltzmann

For a color version of all figures in this chapter, see www.iste.co.uk/akinaga/thermoelectric1.zip.

Thermoelectric Micro/Nano Generators 1,
coordinated by Hiroyuki AKINAGA, Atsuko KOSUGA,
Takao MORI and Gustavo ARDILA.
© ISTE Ltd 2023.

transport theory and high-throughput searches for new TE materials. These efforts have led to the creation of large, open-access computational databases of calculated transport and TE properties. In the last decade, computations have also significantly refined our understanding on the role of defects in TE materials and provided guidance for charge carrier tuning through doping. With the advent of data-driven approaches, TE material development will increasingly employ machine learning and material informatics.

In this chapter, we map the historical contributions of computations in TE material discovery and development, and highlight some recent advances. First, we discuss the general theory of charge carrier and phonon transport, and defect formation and doping. In the following section, we review the application of the theory in computational searches and development of TE materials. Several examples are briefly discussed in this section. The goal is to point the reader to relevant studies without delving into the details. Finally, we highlight several examples of high-throughput searches for new TE materials that use first-principles calculations and/or machine learning models. This chapter provides a broad overview of the computational and data-driven development of TE materials, and we hope that the readers will appreciate and recognize the important role of computations in future TE material development.

2.1. General theory

We review the general theoretical concepts underpinning the physics of TE materials. While concepts such as the TE figure of merit (zT), TE quality factor (β) and Boltzmann transport theory have been summarized elsewhere (Askerov 1994; Lundstrom 2002; Goldsmid 2013, 2017; Koumoto and Mori 2013), we review them here so that this chapter is self-contained. We begin by reviewing Boltzmann transport theory, which is the most commonly used framework for modeling and quantifying transport coefficients that are relevant to TE materials. We discuss the relaxation time approximation and provide brief derivations of the transport coefficients. Optimization of the TE performance critically depends on tuning the electronic carrier concentrations, which is achieved through native or extrinsic doping. In this context, we discuss the standard dilute defect model for calculating defect formation thermodynamics. Alloying is widely used by the TE community to enhance electronic transport, for example, through band convergence, and suppress phonon transport, for example, by increasing phonon–phonon scattering.

We discuss computational methods to model alloys for predicting formation thermodynamics and calculating effective band structures.

2.1.1. *Boltzmann transport theory*

The theory of electronic (and phononic) transport in solids was initially inspired by the kinetic theory of gases, where electrons are treated as classical gaseous molecules. Three fundamental assumptions are imposed in these models: (1) collisions are assumed to instantaneously change the velocity of a particle, (2) interactions with nearby ions and other electrons are neglected during scattering events and (3) scattering events occur with a characteristic probability per unit time, $1/\tau$, where τ is the average time between collisions. These assumptions form the basis of the Drude model (Drude 1900; Ashcroft and Mermin 1976; Kittel and McEuen 1996), in which the electrical current density is directly proportional to the applied field through the electrical conductivity σ of a material, as given by

$$\sigma = \frac{ne^2\tau}{m^*}$$
$$= ne\mu \qquad [2.1]$$

where n is the electronic carrier concentration, e is the fundamental elementary charge, m^* is the effective carrier mass and μ is the carrier mobility.

The model was initially successful in describing electronic transport in metals and explaining empirical relations such as the Wiedemann–Franz law for electronic thermal conductivity (Franz and Wiedemann 1853; Ashcroft and Mermin 1976). Although the "free electron gas" picture was sufficient to describe electronic transport in certain metals, the Drude model has obvious limitations. By the early 1920s, it became clear that electrons are non-classical particles (fermions). Neglecting interactions between electrons ("independent electron" approximation) and between electrons and the periodic potentials of ions in a lattice ("free electron" approximation) are some of the critical issues with the Drude model. To circumvent some of these issues, semi-classical frameworks were developed, in which interactions with the periodic potential of ions in a lattice are treated non-classically, while interactions with external fields are still considered classically (Ashcroft and Mermin 1976; Lundstrom 2002).

The widely used Boltzmann transport theory is based on a semi-classical model, which considers the temporal evolution of the *distribution* of particles $g(\mathbf{r}, \mathbf{k}, t)$ in phase space, instead of tracking the motions of individual particles, for example, charge carriers. Here, $g(\mathbf{r}, \mathbf{k}, t)d\mathbf{r}d\mathbf{k}$ is the number of particles in the volumetric element $d\mathbf{r}d\mathbf{k}$, where $\mathbf{r}$ and $\mathbf{k}$ are the position and momentum (wave) vectors, respectively. While $g(\mathbf{r}, \mathbf{k}, t)$ reduces to the Fermi–Dirac distribution for electronic charge carriers (Fermions) and the Bose–Einstein distribution for phonons (Bosons) under equilibrium conditions, the particle distribution at a point $(\mathbf{r}, \mathbf{k})$ in phase space evolves by the diffusion of particles, driven by external fields, and through scattering under non-equilibrium conditions:

$$\frac{\partial g}{\partial t} = -\underbrace{\mathbf{v}\cdot\frac{\partial g}{\partial \mathbf{r}}}_{\text{Diffusion}} - \underbrace{\frac{\mathbf{F}}{\hbar}\frac{\partial g}{\partial \mathbf{k}}}_{\text{External Field}} + \underbrace{\left(\frac{\partial g}{\partial t}\right)_{\text{scatt}}}_{\text{Scattering}} \quad [2.2]$$

It should be noted that equation [2.2] generally accounts for unsteady state conditions (not to be confused with non-equilibrium conditions), which can occur in the presence of time-varying fields, for example. Under *steady-state* conditions, $\frac{\partial g}{\partial t} = 0$ and the temporal evolution of $g(\mathbf{r}, \mathbf{k}, t)$ arises only due to scattering events into and out of a given state. As a result, the electronic and phonon scattering mechanisms play a key role in determining the transport properties and performance of TE materials.

2.1.2. *Relaxation time approximation*

The distribution evolves as collisions scatter particles into and out of a given state, which is expressed as

$$\left[\frac{\partial g(\mathbf{k})}{\partial t}\right]_{\text{scatt}} = \underbrace{[1 - g(\mathbf{k})]\int\frac{d\mathbf{k}'}{(2\pi)^3}W_{\mathbf{k}',\mathbf{k}}\,g(\mathbf{k}')}_{\text{scattering in}} \\ - \underbrace{g(\mathbf{k})\int\frac{d\mathbf{k}'}{(2\pi)^3}W_{\mathbf{k},\mathbf{k}'}\,[1 - g(\mathbf{k}')]}_{\text{scattering out}} \quad [2.3]$$

where $W_{\mathbf{k},\mathbf{k}'}$ is the transition probability from wave vector $\mathbf{k}$ to $\mathbf{k}'$, and depends on the specific type of scattering event. For example, impurity scattering is described by an interaction term $U(\mathbf{r})$ in a form similar to Fermi's golden rule:

$$W_{\mathbf{k},\mathbf{k}'} = \frac{2\pi}{\hbar} n_i \delta(E(\mathbf{k}) - E(\mathbf{k}'))|\langle \mathbf{k}|U|\mathbf{k}'\rangle|^2 \quad [2.4]$$

The "scattering in" term in equation [2.3] can be understood as a sum over all particles in occupied states $\mathbf{k}'$, which scatter into the state $\mathbf{k}$, weighted by the transition probability $W_{\mathbf{k},\mathbf{k}'}$ and the density of unoccupied states $[1 - g(\mathbf{k})]$. The "scattering out" term can be analogously described as scattering from wave vector $\mathbf{k}$ to $\mathbf{k}'$.

In practice, the nonlinear integro-differential Boltzmann equation (equation [2.2]) is simplified to a linear partial differential equation through the following assumptions: (1) $g(\mathbf{r}, \mathbf{k}, t)$ does not depend on $g(\mathbf{r}, \mathbf{k}, t - dt)$ after a scattering event and (2) $g(\mathbf{r}, \mathbf{k}, t)$ is unaffected by collisions in a region $\mathbf{r}$ with local equilibrium. The two assumptions together form the basis of the relaxation time approximation (RTA), which effectively assumes that scattering processes are described by characteristic relaxation times $\tau_n(\mathbf{r}, \mathbf{k})$ for band index n. By assuming that particles encounter scattering events within a time interval dt with probability $dt/\tau_n(\mathbf{k})$, the scattering term in equation [2.3] becomes,

$$\left[\frac{\partial g(\mathbf{k})}{\partial t}\right]_{\text{scatt}} = \underbrace{\frac{g^0(\mathbf{k})}{\tau_n(\mathbf{k})}}_{\text{scattering in}} - \underbrace{\frac{g(\mathbf{k})}{\tau_n(\mathbf{k})}}_{\text{scattering out}} \quad [2.5]$$

which simplifies the description of transport considerably, while maintaining the idea that the temporal change in the electron distribution function is due to scattering into and out of a given state. We define $g^0(\mathbf{k})$ as the equilibrium distribution function of particles in phase space. By comparing equations [2.3] and [2.5], it becomes clear that RTA effectively assumes that scattering processes are altogether described by a relaxation time $\tau_n(\mathbf{k})$, which itself does *not* depend on the specific shape of $g(\mathbf{r}, \mathbf{k}, t)$.

2.1.3. *Thermoelectric properties*

The transport coefficients, namely the electrical conductivity (σ), Seebeck coefficient (α) and electronic thermal conductivity (κ_e), are understood from the electrical current density and the thermal current density generated by charge carriers, both of which can be expressed in terms of the particle distribution function $g(\mathbf{r}, \mathbf{k}, t)$. The electrical ($\mathbf{J^e}$) and thermal ($\mathbf{J^t}$) current densities can be expressed as,

$$\begin{pmatrix} \mathbf{J^e} \\ \mathbf{J^t} \end{pmatrix} = \sum_n \int \frac{d\mathbf{k}}{4\pi^3} \begin{pmatrix} -e \\ E_n(\mathbf{k}) - \mu \end{pmatrix} \mathbf{v}_n(\mathbf{k}) g_n(\mathbf{k}) \qquad [2.6]$$

where the summation runs over band indices n. In the presence of a temperature gradient and a uniform electric field, the distribution function $g(\mathbf{r}, \mathbf{k}, t)$ under RTA is evaluated as

$$g(\mathbf{k}) = g^0(\mathbf{k}) + \tau(E) \left(-\frac{\partial f}{\partial E} \right) \mathbf{v}(\mathbf{k}) \cdot \left[-e\mathcal{E} - \frac{E(\mathbf{k}) - \mu}{T} \nabla T \right] \qquad [2.7]$$

where $\mathcal{E} = \mathbf{E} + \frac{\nabla \mu}{e}$ is the observed electric field and $\mathbf{E}$ is the applied electric field. Here, f is the Fermi–Dirac distribution and μ is the electron chemical potential. Therefore, the current densities can be expressed as

$$\begin{pmatrix} \mathbf{J^e} \\ \mathbf{J^t} \end{pmatrix} = \begin{pmatrix} \mathbf{L}^{(0)} & -\frac{1}{eT}\mathbf{L}^{(1)} \\ -\frac{1}{e}\mathbf{L}^{(1)} & \frac{1}{e^2 T}\mathbf{L}^{(2)} \end{pmatrix} \begin{pmatrix} \mathcal{E} \\ -\nabla T \end{pmatrix} \qquad [2.8]$$

where,

$$\begin{aligned} \mathbf{L}^{(\alpha)} &\equiv e^2 \int dE \left(-\frac{\partial f}{\partial E} \right) (E - \mu)^\alpha \Sigma(E) \\ \Sigma(E) &\equiv \sum_{n,\mathbf{k}} \tau_n(E) \int \frac{d\mathbf{k}}{4\pi^3} \delta(E - E_n(\mathbf{k})) \mathbf{v}_n(\mathbf{k}) \mathbf{v}_n(\mathbf{k}). \end{aligned} \qquad [2.9]$$

Here, $\mathbf{L}^{(\alpha)}$ represents the generalized transport coefficients and $\Sigma(E)$ is the transport distribution function. The electrical conductivity is the proportionality constant between $\mathbf{J^e}$ and $\mathcal{E}$,

$$\begin{aligned} \sigma &= \mathbf{L}^{(0)} \\ &= e^2 \int dE \left(-\frac{\partial f}{\partial E} \right) \Sigma(E) \end{aligned} \qquad [2.10]$$

The Seebeck coefficient (or thermopower) is the proportionality constant between $\mathcal{E}$ and $-\nabla T$ in an open-circuit, i.e. $\mathbf{J^e} = 0$,

$$\alpha = \frac{1}{eT} \left[\mathbf{L}^{(0)} \right]^{-1} \mathbf{L}^{(1)}$$
$$= \frac{e}{T\sigma} \int dE \left(-\frac{\partial f}{\partial E} \right) (E - \mu) \Sigma(E) \qquad [2.11]$$

Finally, the electronic thermal conductivity is the proportionality constant between $\mathbf{J^t}$ and $-\nabla T$, also in an open-circuit, i.e. $\mathbf{J^e} = 0$,

$$\kappa_e = \frac{1}{e^2 T} \mathbf{L}^{(2)} - \frac{1}{e^2 T} \mathbf{L}^{(1)} \left[\mathbf{L}^{(0)} \right]^{-1} \mathbf{L}^{(1)}$$
$$= \frac{1}{T} \int dE \left(-\frac{\partial f}{\partial E} \right) (E - \mu)^2 \Sigma(E) - \alpha^2 \sigma T \qquad [2.12]$$

The expressions for these transport coefficients can be simplified for a single band (Sofo and Mahan 1994; May and Snyder 2017; Naithani and Dasgupta 2019; de Boor 2021; Zhu et al. 2021a; Toriyama et al. 2022b) and are often used to identify the dominant scattering mechanism(s) from experimental measurements (Kang et al. 2018).

The lattice contribution to the thermal conductivity, termed lattice thermal conductivity (κ_L), can be derived from the thermal current density and the phonon distribution function $n_m(\mathbf{r}, \mathbf{q}, t)$, where m is the phonon branch index and $\mathbf{q}$ is the phonon wave vector. Since phonons diffuse and scatter, much like electronic charge carriers, the equivalent transport equation for phonons, known as the Peierls–Boltzmann transport equation, in steady state is given by

$$\left(\frac{\partial n}{\partial t} \right)_{\text{scatt}} = \mathbf{v}_m(\mathbf{q}) \cdot \nabla T \frac{\partial n}{\partial T} \qquad [2.13]$$

where $\mathbf{v}_m(\mathbf{q}) = \frac{\partial \omega_m(\mathbf{q})}{\partial \mathbf{q}}$ is the group velocity of the phonon mode m at wave vector $\mathbf{q}$. The term on the left-hand side has contributions from various phonon scattering mechanisms. We solve for the non-equilibrium distribution function $n_m(\mathbf{r}, \mathbf{q}, t)$ to obtain the phonon lifetimes, which are subsequently used to calculate κ_L:

$$\kappa_{L,\alpha\beta} = \sum_m \int_0^{\omega_{\max}} d\omega C_m(\omega) v_{m,\alpha}(\omega) v_{m,\beta}(\omega) \tau_m(\omega) \qquad [2.14]$$

where $C_m(\omega)$ is the mode-specific heat capacity, $v_{m,\alpha}(\omega) = \frac{d\omega}{d\mathbf{q}}$ is the phonon group velocity in the α Cartesian direction, and $\tau_m(\omega)$ is the phonon relaxation time. Equation [2.14] has a form similar to equations [2.10]–[2.12], in which an integral over phonon modes (vs. electronic states) is performed.

2.1.4. Defect theory

Often, the TE performance of a material needs to be optimized by tuning the electronic carrier concentration (electrons or holes) through doping. While the electronic and phonon transport properties shed light on the potential for TE performance, the actual achievable zT is governed by the ability to dope a material with the desired electronic carriers (electrons or holes) to an optimal concentration. The thermodynamics of defect formation, including native defects and dopants, enable or limit the tuning of the electronic carrier concentration.

The doping efficacy depends, first and foremost, on the nature and formation energetics of the native defects. In TE materials, it is often the point defects that dictate the doping properties. Point defects include vacancies, interstitials, antisites and even small defect complexes. First-principles computational modeling of defect formation in materials has significantly refined our understanding on the role of defects in doping. More importantly, predictive defect modeling has provided guidance for choosing appropriate dopants. Here, we briefly discuss the theory behind point defect modeling in semiconductors using the supercell approach (Lany and Zunger 2008).

Within the supercell approach, the defect formation energy $\Delta E_{D,q}$ (D is the defect type, q is the charge state) is calculated from the total energies as:

$$\Delta E_{D,q} = E_{D,q} - E_{\text{host}} + \sum_i +n_i\mu_i + qE_F + E_{\text{corr}} \qquad [2.15]$$

where $E_{D,q}$ and E_{host} are the total energies of the supercell with and without the defect, respectively. E_F is the Fermi energy, which is typically referenced to the valence band maximum (VBM), i.e. $E_F = 0$ eV at VBM. The chemical potential μ_i of element i is expressed relative to a reference state (μ_i^0) such that $\mu_i = \mu_i^0 + \Delta\mu_i$. A certain number of atoms (n_i) of element i are added ($n_i < 0$) or removed ($n_i > 0$) from the host supercell to form the defect D. Since $\Delta E_{D,q}$ is calculated using *periodic* supercells, the defect formation energy needs to be "corrected" for the finite-size effects. These corrections are denoted by E_{corr}. We refer the reader to Lany and Zunger (2008) and Freysoldt et al. (2014) for a detailed discussion on the finite-size corrections.

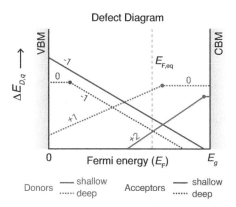

Figure 2.1. *Schematic of a defect diagram. Calculated defect formation energies ($\Delta E_{D,q}$) are plotted against the Fermi energy (E_F) for various defects and dopants, which are denoted by straight lines. In semiconductors, E_F is typically referenced to the valence band maximum (VBM) and plotted between the VBM and conduction band minimum (CBM). The condition of charge neutrality determines the position of the equilibrium (or pinned) E_F, which is a function of temperature. The formation energy of defects at the E_F is relevant*

The calculated defect formation energies are graphically presented as a "defect diagram", as shown schematically in Figure 2.1, where the defect formation energy ($\Delta E_{D,q}$) is plotted against the Fermi energy (E_F). For a semiconductor with a band gap, E_F is typically referenced to the valence band maximum (VBM), i.e. $E_F = 0$ eV at VBM, and plotted between the VBM and the conduction band minimum (CBM). Each defect is represented by a set of straight lines, where the slope of the line is the charge state q of the defect at a given E_F. The linear dependence of $\Delta E_{D,q}$ on E_F is evident from equation [2.15]. Conventionally, only the lowest-energy (favorable) charge state of a defect at a specific E_F is drawn because that charge state has the highest concentration; however, non-zero concentrations of other charge states exist, as determined by Boltzmann statistics. Neutral defects are denoted by horizontal lines, while donors (acceptors) are denoted by lines with positive (negative) slopes.

For a given defect, the E_F at which the favorable charge state changes are the charge transition levels (CTLs), which correspond to quasi particle defect levels. Donor and acceptor defects are further classified as shallow or deep depending on the location of the CTLs. If there are no CTLs inside the band gap or the CTL lies close to the relevant band edges (donor close to conduction band, acceptor close to valence band), typically within a few $k_\mathrm{B}T$ (k_B is Boltzmann constant, T is temperature), the defect is considered a shallow defect. The charge carrier associated with the shallow defect level can be easily thermally excited into the bands, thereby creating free charge carriers. In contrast, deep defects are characterized by defect levels far from the relevant band edges. The associated charge carriers are bound to such deep defects and not available as free carriers.

We can calculate the defect and charge carrier concentrations self-consistently from these defect diagrams. The positively (donor) and negatively (acceptor) charged defects and free carriers (electrons, holes) must satisfy charge neutrality. The free charge carrier concentrations can be calculated self-consistently by solving for the equilibrium Fermi energy (E_F^{eq}) satisfying the charge neutrality condition,

$$\sum_D q N_\mathrm{D} e^{-\Delta E_{\mathrm{D},q}/k_\mathrm{B}T} + p - n = 0 \qquad [2.16]$$

where the sum runs over all defects D, and N_D is the site concentration where defect D can be formed. The equilibrium Fermi energy (E_F^{eq}) generally lies around the intersection of the lowest $\Delta E_{\mathrm{D},q}$ acceptor and donor defects. The electrical conductivity type (p-type or n-type) is determined by the position of E_F^{eq}. The closer E_F^{eq} is to the corresponding band edge, the higher the concentration of the charge carriers is. For example, if E_F^{eq} is close to the conduction band, electrons are the majority charge carriers. The theoretical framework described above can be extended to understand extrinsic doping.

2.2. Applications

2.2.1. *Transport calculations*

2.2.1.1. *Charge carrier transport*

First-principles modeling and prediction of electronic transport properties in materials typically involve two steps: (1) calculate the electronic structure of the material and (2) use Boltzmann transport theory to calculate the electronic transport coefficients. Several software packages are available to solve Boltzmann transport equations, where the electronic structure and carrier relaxation times (or an approximation) are provided as inputs. These softwares include BoltzTraP (Madsen and Singh 2006; Madsen et al. 2018), LanTraP (Wang et al. 2018b), BoltzWann (Pizzi et al. 2014), Perturbo (Zhou et al. 2021), EPIC STAR (Deng et al. 2020), AMSET (Faghaninia et al. 2015; Ganose et al. 2021), TransOpt (Li et al. 2021a) and ELECTRA (Li et al. 2021b). The availability of these software packages and detailed documentations have enabled calculations of electronic transport coefficients to become a routine procedure. While the electronic structure is straightforwardly calculated, the determination of carrier relaxation times from first principles is computationally challenging. Consequently, various approximations are employed to describe the relaxation time.

The constant relaxation time approximation (CRTA) is most commonly used to calculate the electronic transport coefficients. In CRTA, the relaxation time τ is assumed to be an energy-independent constant in the entire Brillouin zone. This approximation was initially justified on the grounds that the Seebeck coefficient can be approximately treated as independent of τ (Schulz et al. 1992; Madsen and Singh 2006). Predictions based on CRTA are in reasonable agreement with experiments for some materials, including Bi_2Te_3 (Scheidemantel et al. 2003), lead chalcogenides (Wang et al. 2007; Xu et al. 2010, 2011; Singh 2010; Parker et al. 2012; Parker and Singh 2014), half-Heuslers (Yang et al. 2008; Fang et al. 2016), diamond-like semiconductors (Sevik and Çağın 2009) and zinc antimonides (Bjerg et al. 2011).

Initially, transport calculations with CRTA provided computational guidance for optimizing the performance of known TE materials. Notably, PbSe was predicted to be a good p-type TE material through transport calculations employing CRTA (Parker and Singh 2010), which was later confirmed experimentally by heavily doping with Na (Wang et al. 2011) and alloying with SrSe (Wang et al. 2014). This further inspired transport calculations for optimizing n-type PbSe (Parker et al. 2012), which was later demonstrated experimentally through resonant Al doping (Zhang et al. 2012a; Lee et al. 2014b). The TE properties of the ternary antimonide CoSbS were investigated with CRTA transport calculations, (Parker et al. 2013) and experimentally realized through Ni doping (Liu et al. 2015; You et al. 2018). A zT of ~0.5 at 900 K was achieved in Ni-doped samples, a significant increase from the modest zT of ~0.2 in undoped CoSbS.

As CRTA is implemented in several software packages (as noted above) and is computationally economical, it has been extensively used in several high-throughput TE material searches and resulted in the creation of large databases of electronic transport calculations (Ricci et al. 2017; Choudhary et al. 2020a). Madsen (2006) is credited with some of the earliest attempts at using computations to guide the search for new TE materials. The study employed Boltzmann transport theory with CRTA to investigate a large set of antimonide compounds; LiZnSb was predicted to be a promising n-type TE material (Madsen 2006). Unfortunately, it was found to be experimentally challenging to realize n-type LiZnSb (Toberer et al. 2009). We now understand that LiZnSb is a degenerate p-type semiconductor, and cannot be doped n-type, at least under equilibrium conditions (Gorai et al. 2019). Since then, similar high-throughput TE material searches have employed the CRTA (Opahle et al. 2013; Zhu et al. 2015; Miyata et al. 2018). For example, a search among XYZ_2 compounds identified TmAgTe$_2$ as a promising TE material (Zhu et al. 2015) and later inspired studies of YCuTe$_2$ (Aydemir et al. 2016; Pöhls et al. 2018), RECuTe$_2$ (RE = Tb, Dy, Ho, Er) (Lin et al. 2017) and TmAg$_x$Cu$_{1-x}$Te$_2$ alloys (Bai et al. 2021). A high-throughput search among 48,000 compounds revealed that some metal phosphides exhibit good TE performance (Pöhls et al. 2017). One of these compounds, cubic NiP$_2$, was synthesized and the measured Seebeck coefficient was found to be in agreement with the predicted values in a wide temperature range (Pöhls et al. 2017). In another study, thousands of compounds with high power factors were identified and documented in the NIST-JARVIS database (Choudhary et al. 2020a, 2020b).

Charge carrier transport calculations have also revealed composition–structure–property relationships that provide useful guidance in the discovery and design of new TE materials. For example, a high-throughput study showed that oxide compounds tend to have a lower power factor compared to chalcogenide (S, Se, Te) compounds, and therefore, it is more difficult to find high-performing oxide TE materials (Chen et al. 2016). The electronic fitness function (EFF) has been proposed as a metric to quantify the suitability of the electronic structure to achieve good TE performance (see Figure 2.2) (Xing et al. 2017; Feng et al. 2019). The EFF is defined as

$$\text{EFF} = \frac{\sigma}{\tau} \frac{S^2}{N^{2/3}} \quad [2.17]$$

where τ is the scattering time and N is the volumetric density of states. EFF addresses the competing relationship between the electrical conductivity

σ and the thermopower S arising from, for example, band convergence and multiple electronic valleys (Xing et al. 2017; Feng et al. 2019). Therefore, the EFF ranks materials with high valley degeneracy higher, as multiple carrier pockets lead to high S while simultaneously maintaining low effective carrier mass. A high-throughput screening using EFF as a descriptor of the electronic transport identified Zintl phases Na_2AuBi (p-type) and KSnSb (n-type) as promising materials with high power factors (Xing et al. 2017), owing to high band degeneracy (see Figure 2.2). In a similar study using EFF for screening half-Heuslers, LiBSi was identified as a promising TE material (Feng et al. 2019).

Recent works that use the CRTA have also elucidated the quantitative implications of non-parabolic band structures on the electrical transport properties. The Fermi surfaces of the valence bands in rock-salt lead chalcogenides exhibit a tubular geometry (Parker et al. 2013; Brod and Snyder 2021), akin to quantum-well structures (Hicks and Dresselhaus 1993) and in $SrTiO_3$ (Tsuda et al. 2013; Sun and Singh 2016; Dylla et al. 2019). Transport calculations employing CRTA suggested that the TE performance of n-type GeTe and SnTe, for which the conduction band structures deviate considerably from the conventional parabolic shape (Toriyama et al. 2022a), may exceed those of the lead chalcogenides due to the Fermi surface geometry (Chen et al. 2013). This later sparked ideas to engineer low-dimensional transport to improve the TE performance of full-Heusler compounds (Bilc et al. 2015).

Despite the computational efficiency of CRTA, there are serious shortcomings of this approximation. While thermopower is independent of τ in CRTA, the magnitude of τ still affects the electrical conductivity and the electronic contribution to the thermal conductivity. This effectively restricts the constant to a mere fitting parameter that must be set on a case-by-case basis to obtain useful predictions. In fact, Chen et al. (2016) found that invoking a "universal scattering time" to predict the power factors of a large set of compounds resulted in poor agreement with experimental results, with a fitting quality of $r^2 = 0.33$. Katsura et al. (2019) found that τ varies by at least two orders of magnitude by analyzing hundreds of published data on n-type PbTe. Ganose et al. (2021) found that better agreement with experimentally measured carrier mobilities can be achieved by using density functional perturbation theory (Baroni et al. 2001) with Wannier interpolation (Marzari and Vanderbilt 1997; Souza et al. 2001; Marzari et al. 2012; Verdi and Giustino 2015). Xu et al. (2020) showed that a variational approach can be used to obtain

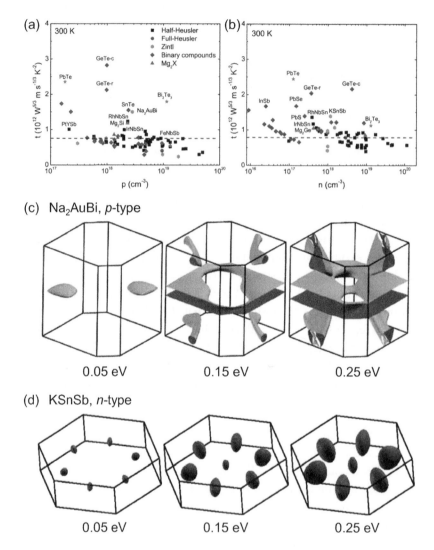

Figure 2.2. The electronic fitness function (EFF) is a metric to quantify the features of the electronic structure that promote favorable charge carrier transport for thermoelectrics. (a–b) Calculated EFF of various materials and the corresponding optimal hole and electron carrier concentrations. Materials that lie above the horizontal dashed line are considered promising candidates for thermoelectric applications. Constant-energy Fermi surfaces at various isosurface levels of (c) the valence band of Na_2AuBi, which has a high optimal p-type EFF. (d) The conduction band of KSnSb, which has a high optimal n-type EFF. Reproduced from Xing et al. (2017) with permission from the American Physical Society

better agreement with measured Seebeck coefficients of metals. Graziosi et al. (2019) demonstrated that considering electron–phonon interactions yields a different power factor ranking of half-Heusler compounds than with CRTA. These results suggest that overlooking the details of the scattering physics may misguide high-throughput searches for TE materials. While the CRTA offers a computationally efficient method for calculating charge carrier transport properties, more rigorous approaches which account for various scattering mechanisms have been introduced since.

The four primary electron–phonon scattering mechanisms are (Yu and Cardona 2010):

1) acoustic phonon deformation potential scattering, where electrons are scattered nearly elastically by short-range interactions with long-wavelength longitudinal acoustic phonons;

2) acoustic phonon piezoelectric scattering, where the electric polarization induced by mechanical stress in non-centrosymmetric crystals scatters electrons via long-range Coulombic interactions;

3) non-polar optical phonon scattering, where the electronic structure is perturbed by bond length fluctuations induced by optical phonons;

4) polar optical phonon scattering (Frölich mechanism), in which long-wavelength longitudinal optical phonons induce a macroscopic electric field by uniformly displacing ions of opposite charge.

Mechanisms (1) and (3) are both deformation potential mechanisms, where changes in either the volume of the unit cell (acoustic phonons) or the interatomic bond lengths (optical phonons) create deformation potentials that scatter charge carriers. Mechanisms (2) and (4) both involve an electric polarization induced by phonons.

Explicit treatment of electron–phonon interactions in materials is now available in softwares that can directly compute scattering rates from first principles (Bernardi 2016; Giustino 2017). EPW (Giustino et al. 2007; Noffsinger et al. 2010; Margine and Giustino 2013; Poncé et al. 2016) is often used in conjunction with the density functional theory (DFT) code Quantum Espresso (QE) to calculate the electron–phonon coupling matrix and charge transport properties such as carrier mobility. The calculations involve density functional perturbation theory (Baroni et al. 2001) and Wannier interpolation (Marzari and Vanderbilt 1997; Marzari et al. 2012) of the electronic structure.

A number of studies have been dedicated to modeling the electron–phonon interactions in PbTe. While it is traditionally believed that acoustic phonon scattering is the dominant charge scattering mechanism in PbTe due to the T^r ($r \leq -3/2$) temperature dependence of the mobility at carrier concentrations of $\sim 10^{19}$ cm^{-3} (Pei et al. 2012, 2014; Ravich 2013), first-principles calculations suggest that acoustic phonon scattering is dominant only at high carrier concentrations of $\sim 10^{20}$ cm^{-3} (Cao et al. 2018). At carrier concentrations considered in experiments, it is found that scattering by longitudinal optical phonons is the dominant mechanism for both electrons (Song et al. 2017; Cao et al. 2018) and holes (D'Souza et al. 2020) (see Figure 2.3). Cao et al. (2018) and Song et al. (2017) independently showed that charge transport in n-type PbTe is primarily limited by longitudinal optical phonons unless the carrier concentrations are high enough to sufficiently screen the optical phonons (see Figure 2.3(a) and (b)). D'Souza et al. (2020) found that although scattering between the L and Σ bands by longitudinal optical phonons decreases the electrical conductivity in p-type PbTe, the overall TE performance is enhanced since the scattering increases the thermopower and decreases the electronic thermal conductivity (see Figure 2.3(c)). Similarly, Zhou et al. (2018) found that the acoustic phonon deformation potentials of half-Heusler compounds tend to be lower than optical phonon deformation potentials, giving rise to their high power factors. The incorrect attribution to acoustic phonon scattering may partly be due to the similar temperature dependencies of the carrier mobility to polar optical phonon scattering. This also indicates that the acoustic phonon deformation potential scattering mechanism may not be the dominant mechanism for many other polar materials reported in the literature, including GaAs (Sjakste et al. 2015; Zhou and Bernardi 2016; Liu et al. 2017), ZnX (X = S, Se) (Ding et al. 2021), SnSe (Ma et al. 2018), CsPbBr$_3$ (Zhou and Zhang 2020) and full-Heusler compounds Ba$_2$AuBi, Sr$_2$AuBi and Sr$_2$AuSb (Park et al. 2019, 2020).

Such detailed electron–phonon interaction calculations have been used to identify potential TE materials. One such example is n-type Ba$_2$AuBi, which is a full-Heusler compound predicted to reach zT = 5 at 800 K (Park et al. 2019; Ma et al. 2020). The rather large disparity in the predicted zT for different carrier types ($zT_{\max} \approx 5$ for n-type, $zT_{\max} \approx 2-3$ for p-type) has been suggested to be in part due to differences in the acoustic deformation potential scattering, in which highly dispersive conduction bands limit the scattering phase space and weaken the acoustic phonon scattering of electrons, whereas the relatively flatter valence bands limit the hole lifetimes by increasing the scattering phase space. It is worth reiterating that charge carrier scattering by polar optical phonons has been suggested to be the dominant electron–phonon scattering mechanism in Ba_2AuBi, reducing the electron lifetimes by nearly an order of magnitude at the conduction band edges (Park et al. 2019).

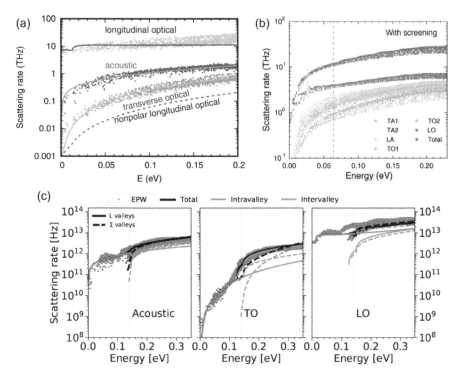

Figure 2.3. *Charge carrier scattering rates from electron–phonon interactions in (a–b) n-type and (c) p-type PbTe. Scattering rates are calculated at 300 K. Scattering by longitudinal optical phonons is found to be the dominant scattering mechanism. Reproduced from Song et al. (2017), Cao et al. (2018) and D'Souza et al. (2020) with permission from the American Physical Society and Elsevier*

Since the rigorous calculation of electron–phonon interactions is computationally expensive, approximations that are more accurate than CRTA have been proposed to calculate charge carrier transport properties. One such alternative is the constant electron–phonon coupling approximation, where the scattering time depends on a constant deformation potential as

$$\frac{1}{\tau_{n,\mathbf{k}}} = \frac{2\pi k_B T E_{\text{def}}^2}{V \hbar G} \sum_{m,\mathbf{k}'} \delta\left(\epsilon_{n,\mathbf{k}} - \epsilon_{m,\mathbf{k}'}\right)$$ [2.18]

where E_{def} is the deformation potential of the band edge (Li et al. 2021a). This method is implemented in the TransOpt package (Li et al. 2021a). Several high-throughput screening studies have employed this method, which notably resulted in the discovery of $Cd_2Cu_3In_3Te_8$ with peak $zT > 1$ (Xi et al. 2018). In this study, Xi et al. found that vacancy-containing diamond-like semiconductors with 1-2-4 stoichiometry possess relatively high power factors (see Figure 2.4(a)), from which the Cu-intercalated $Cd_2Cu_3In_3Te_8$ was discovered (see Figure 2.4(b)). Studies leveraging cation disorder (Pan et al. 2019) and embedded nanodomains (Pan et al. 2021) have further improved the TE performance to $zT = 1.2$. Other studies have investigated the TE performance of similar compounds such as $Zn_2Cu_3In_3Te_8$ (Zhang et al. 2021). Using the constant electron–phonon coupling approximation, Li et al. (2019) predicted 24 diamond-like ABX_2 compounds that exhibit promising TE performance, citing that pnictide-based ABX_2 compounds tend to have higher power factors than chalcogenide-based compounds. The TE performance of one of the predicted candidates, $AgInSe_2$, was investigated experimentally (Vasiliev et al. 2021).

Another alternative to explicitly calculating electron–phonon interactions is the ab initio model for calculating mobility and Seebeck coefficient using the Boltzmann transport (aMoBT) equation (Faghaninia et al. 2015), which is implemented in the AMSET software (Ganose et al. 2021). Instead of assuming a constant relaxation time, the momentum relaxation time approximation is used to calculate the rates of different charge carrier scattering mechanisms, resulting in better predictions of mobility and Seebeck coefficient compared to CRTA (Faghaninia et al. 2015). Pöhls et al. found that the electrical resistivity of cubic NiP_2 calculated by the aMoBT model better matches the measured resistivity. Using this method, several compounds in the bournonite structure were suggested as potential candidates for TE applications (Faghaninia et al. 2017). A high-throughput search using the implementation in the AMSET package identified rare-earth phosphides, $RECuZnP_2$, as promising TE materials (Pöhls et al. 2021). A maximum zT of 0.5 at 800 K was reached in experiments for some of the suggested compounds (see Figure 2.4(c) and 2.4(d)). Other materials that have been predicted to be good TE materials include metal halides (Jung et al. 2021), binary Zintl phases (Kumar and Bera 2021) and mixed-anion compounds (Rahim et al. 2021).

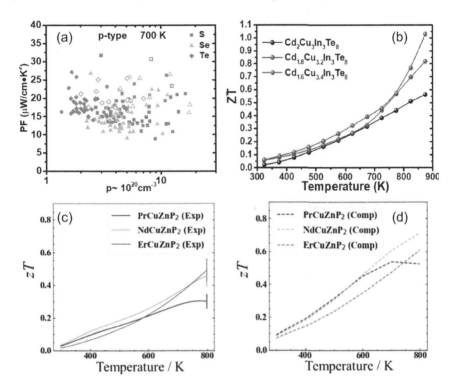

Figure 2.4. *(a) Calculated power factors as a function of hole concentration for a set of chalcogenide-based diamond-like semiconductors using the constant electron–phonon coupling approximation. (b) The search identified $Cd_2Cu_3In_3Te_8$-based thermoelectrics to exhibit high zT. Screening of 20,000 ICSD compounds identified $RECuZnP_2$ (RE = Pr, Nd, Er) compounds as promising thermoelectric materials. The measured (c) and predicted (d) zT are shown for the three $RECuZnP_2$ compounds. Reproduced from Xi et al. (2018) and Pöhls et al. (2021) with permission from the American Chemical Society and the Royal Society of Chemistry*

2.2.1.2. Phonon transport

Phonon frequencies and lifetimes in a material are typically calculated using some set of approximations that balance the computational expense and accuracy (Xia et al. 2020). These different approximations arise from the Taylor expansion of the interatomic potential with respect to atomic displacements, which are calculated using, for example, density functional perturbation theory (DFPT) with linear response approaches (Lindsay et al. 2019). While phonon dispersion is described by harmonic force constants, the interaction between phonons themselves leads to anharmonicity, which requires an explicit treatment of three-phonon, and sometimes even four-phonon, processes to accurately predict the lattice thermal conductivity (κ_L) of a material (Lindsay et al. 2019; Xia et al. 2020). However, it is challenging to calculate higher-order anharmonicity in materials with large unit cells with DFPT because of the dramatic increase in the computational expense. To address these issues, the compressive sensing lattice dynamics (CSLD) approach has been used to calculate only the dominant anharmonic terms in the Taylor series of the total energy (Zhou et al. 2014, 2019a, 2019b).

A number of user-friendly software packages have implemented methods for calculating phonon transport properties from first-principles calculations. These software include ShengBTE (Li et al. 2014), almaBTE (Carrete et al. 2017), phono3py (Togo et al. 2015), TDEP (Hellman et al. 2011; Hellman and Abrikosov 2013; Hellman et al. 2013) a-TDEP (Bottin et al. 2020), FourPhonon (Han et al. 2022), AICON (Fan and Oganov 2020, 2021), PhonTS (Chernatynskiy and Phillpot 2015) and ALAMODE (Tadano et al. 2014). The availability of these software has made direct calculation of phonon transport properties more accessible, and as a result, there has been a surge in the number of studies dedicated to investigating thermal transport in TE materials.

Several studies have focused on searching for materials with low κ_L. Yang et al. (2018) predicted that cubic Li$_3$Sb has a low κ_L of 2.2 W/mK at room temperature from explicit calculations of third-order interatomic force constants. Yahyaoglu et al. (2021) synthesized cubic Li$_3$Sb and measured κ_L ~2.7 W/mK at 325 K. While a zT of 0.4 was obtained experimentally, a single parabolic band model showed that the TE performance can be optimized up to 0.8 (Yahyaoglu et al. 2021), verifying that cubic Li$_3$Sb is a promising TE material (Yang et al. 2018).

Large anharmonicities often lead to increased phonon–phonon scattering and low lattice thermal conductivity in crystalline solids. The anharmonicity strength is quantified by the Grüneisen parameter γ_i, which is defined by how the phonon mode frequencies ($\omega_{i,q}$) change with the volume of the unit cell (V):

$$\gamma_{i,q} = -\frac{\partial \ln(\omega_{i,q})}{\partial ln(V)} \quad [2.19]$$

where i is the mode index and q is the phonon wave vector. Materials with low κ_L such as $CsAg_5Te_3$ (Lin et al. 2016), half-Heusler CoNbSi (Ye et al. 2021) and $AgBi_3S_5$ (Tan et al. 2017) exhibit large anharmonicity as evidenced by their large Grüneisen parameters. Resonant doping has also been suggested to increase anharmonicity in materials. In such materials, highly delocalized electron densities lead to lower transverse optical mode frequencies (Lee et al. 2014a). An accurate treatment of four-phonon processes is sometimes required to explain the low κ_L in highly anharmonic materials, as shown for PbTe (Xia 2018) and BAs (Feng et al. 2017). In PbTe, four-phonon scattering processes are responsible for the significant reduction in the lattice thermal conductivity. Inclusion of this higher-order phonon interaction is necessary to match the experimentally measured thermal expansion and thermal conductivity (Xia 2018).

Stereochemically active lone pair electrons have also been found to contribute to higher anharmonicity by increasing electrostatic repulsion and bonding asymmetry (Skoug and Morelli 2011; Nielsen et al. 2013; Lai et al. 2015). For example, He et al. (2017) proposed that Bi_2PdO_4 is a promising TE material where the $6s^2$ Bi lone-pair electrons induce strong bond anharmonicity (see Figure 2.5(a)). An experimental study later confirmed the predicted lattice thermal conductivity, although the power factor of undoped Bi_2PdO_4 was found to be low, yielding a maximum zT of 4×10^{-4} at 680 K (Kayser et al. 2020). A combined computational and experimental study also revealed that the lone-pair electrons of Sb in $CuSbS_2$ play a critical role in lowering κ_L, as evidenced by the κ_L, which is lower than expected from a simple alloy model of $CuSb_{1-x}Ga_xS_2$ (Du et al. 2017). The lower κ_L of Cu_3SbSe_3 compared to that of Cu_3SbSe_4 (Skoug et al. 2010) was attributed to the lone pair of electrons on Sb that slightly distorts the trigonal Sb–Se bonding in Cu_3SbSe_3, as opposed to the tetrahedral Sb–Se bonding in Cu_3SbSe_4 (Skoug and Morelli 2011; Zhang et al. 2012b). It was found that the low κ_L of $Ba_6Sn_6Se_{13}$ is partly due to the lone-pair electrons of Sn^{2+}, which contribute to the phonon density of states at low frequencies (Gunatilleke et al. 2021). However, it should be noted that lone-pair electrons do not necessarily lead to lower κ_L, but the effect rather depends on the coordination environment of the atom with the lone pair (Wang et al. 2018a).

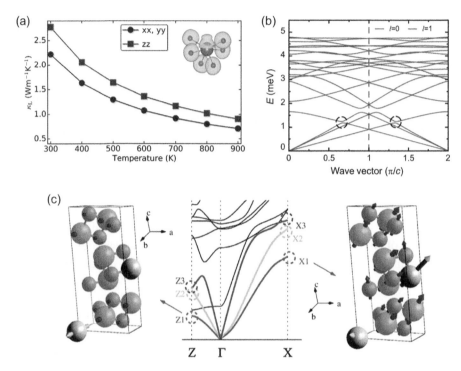

Figure 2.5. *Phonon scattering mechanisms that lower κ_L. (a) Bi lone-pair electrons are responsible for the low κ_L of Bi_2PdO_4. (b) Avoided crossings (circled) in $Ba_6Sn_6Se_{13}$ are understood from the symmetries of the phonon modes, which prohibit band crossings leading to low group velocities and increased scattering. (c) Double rattling mechanism in $AgBi_3S_5$ at the Z and X points of the Brillouin zone. Reproduced from He et al. (2017), Tan et al. (2017) and Gunatilleke et al. (2021) with permission from the American Chemical Society and the American Physical Society*

Rattling modes can also lead to low lattice thermal conductivity. Rattling phonon modes are indicated by avoided crossings between the strongly coupled acoustic and optical branches, resulting in a lower phonon group velocity and increased scattering (He et al. 2016). Avoided crossings are understood in terms of the phonon symmetries. Gunatilleke et al. (2021) showed that, in $Ba_6Sn_6Se_{13}$, the chiral and screw symmetries of the crystal structure prohibit the crossing of phonon bands with the same quasiangular momentum (see Figure 2.5(b)). In the half-Heusler structure, a weakly bounded filler atom has been found to cause an avoided crossing of the longitudinal acoustic and optical phonon branches (Feng et al. 2020). Due to avoided crossings observed in phonon dispersion, a "concerted" rattling mechanism of Ag was proposed as the cause for the low lattice thermal conductivity of $CsAg_5Te_3$ (Lin et al. 2016). In contrast to the vibration of a sublattice, a different kind of rattling mechanism is proposed to explain the low thermal conductivity of $AgBi_3S_5$, in which only a subset of the Ag and Bi sublattices vibrate – a mechanism named "double rattling" (see Figure 2.5(c)) (Tan et al. 2017).

To facilitate the fast screening of materials with low (or high) κ_L, models based on classical theory have been developed. Yan et al. (2015) developed a semi-empirical model for κ_L, where the model inputs are obtained from inexpensive DFT calculations. Subsequently, Miller et al. (2017b) improved upon the model to incorporate the effects of anharmonicity through a modeled Grüneisen parameter that depends only on the average coordination number of atoms in the structure. These fitted models enable fast and large-scale computational predictions of κ_L. A model to predict κ_L of zinc antimonide compounds was developed by Bjerg et al. (2014), where the model inputs can be obtained from the phonon dispersion alone. The low κ_L of zinc antimonides was attributed to the Zn and Sb atoms forming low-lying optical modes, in contrast to the previous understanding that rattling motion of the Sb–Sb dumbbells contribute to the low-lying optical modes (Schweika et al. 2007). Moreover, the low κ_L was attributed to anharmonic motions of Sb atoms bonded to Zn. It is interesting to note, however, that this model proposed by Bjerg et al. (2014) overestimates κ_L in a more diverse set of compounds (Garrity 2016). Nevertheless, the model was also used to screen for low-κ_L materials and found perovskites to exhibit strong anharmonicity and low κ_L.

Point defect scattering of phonons, which is another mechanism by which κ_L is lowered, has been studied with first-principles computations. In the diamond-like semiconductor, $AgGaTe_2$, it was predicted that Ag vacancies and In_{Ga} substitutional defects lower κ_L considerably due to lattice distortions induced by the defects and the rattling behavior of Ag vacancies (Zhong et al. 2021). This reduction was demonstrated experimentally by comparing $Ag_{0.85}Ga_{0.85}In_{0.15}Te_2$ (κ_L = 0.08 W/mK at 850 K) to $AgGa_{0.93}Te_2$ (κ_L = 0.18 W/mK), leading to an overall improvement in zT from 1.05 to 1.44 at 850 K (Zhong et al. 2021). A similar idea was explored for half-Heusler compounds, which despite their exceptional electrical transport properties suffer from high κ_L. Anand et al. (2019) considered combining the non-valence-balanced TiFeSb (17 electrons) and TiNiSb (19 electrons) half-Heusler compounds into the valence-balanced Ti$_2$FeNiSb$_2$, which was predicted to be stable and possess low κ_L. The compound was experimentally synthesized and shown to have κ_L almost three times lower than that of TiCoSb (Anand et al. 2019) due to Fe/Ni disorder (Liu et al. 2019).

2.2.2. Defect and doping calculations

A thermoelectric material needs to be doped with a specific carrier type (electrons or holes) and to a desired carrier concentration to optimize the zT (Snyder and Toberer 2008). Therefore, it is crucial to understand and quantitatively model the role of native defects and extrinsic dopants. The development of the modern theory of defects in semiconductors has made it possible to predictively model the defect formation thermodynamics (Anand et al. 2022) using first-principles calculations. In recent years, several software packages for calculation and visualization (see Figure 2.1) have been developed, including Pylada (Goyal et al. 2017a), PyCDT (Broberg et al. 2018), PyDEF (Péan et al. 2017; Stoliaroff et al. 2018), MAST (Mayeshiba et al. 2017), CoFFEE (Naik and Jain 2018), Spinney (Arrigoni and Madsen 2021) and DASP (Huang et al. 2022). With the availability of such computational tools and methods, first-principles defect modeling is being increasingly used in the discovery and design of existing and new thermoelectric materials.

A quantitative assessment of the defect formation thermodynamics (and, as a result, dopability) of semiconductors is contingent on a variety of computational uncertainties. Aside from the choice of DFT functional (Peng et al. 2013), uncertainties can arise from spurious interactions due to periodic boundary conditions, band filling effects of shallow defects and alignment issues of the average electrostatic potential (Burstein 1954; Lany and Zunger 2008). A major obstacle in modeling defect energetics in TE materials particularly is the prediction of band edge positions relative to vacuum, which influence the dopability and intrinsic carrier concentration of a material. West et al. (2012), for example, showed that spin–orbit coupling influences the predicted conductivity type of Bi$_2$Se$_3$ due to a shifting of the absolute band edge positions. Goyal et al. (2017b), on the other hand, studied the critical role of the band edge positions in PbTe. Due to the narrow band gap and non-negligible spin–orbit coupling effects, the study found that a combination of the hybrid HSE functional, spin–orbit coupling and GW quasiparticle calculations are necessary to accurately capture the intrinsic conductivity type and measured carrier concentration (Goyal et al. 2017b).

In many cases, it is difficult, or even nearly impossible, to dope a material to its optimal zT, despite predictions of high TE performance from electrical transport modeling. The dopability of a material can be understood from native defect energetics, since intrinsic defects that readily form may compensate for charge carriers generated by dopants. For example, high zT was initially predicted for n-type LiZnSb, partly due to the high conduction band degeneracy (Madsen 2006). However, the material was found to be p-type (Toberer et al. 2009) due to the low formation energy of acceptor-like cation vacancies (Gorai et al. 2019), prohibiting electrons from being the dominant charge carrier. Mg_3Sb_2 is another material where n-type conductivity was initially difficult to synthesize. Historically, Mg_3Sb_2 was almost exclusively synthesized as a p-type material due to the high volatility of Mg and facile formation of acceptor-like Mg vacancies (see Figure 2.6(a) and (b)). It was only until recently that, with guidance from defect calculations, n-type Mg_3Sb_2 was realized with Te-doping under Mg-rich growth conditions (see Figure 2.6(c) and (d)) (Ohno et al. 2018). It was also suggested that La-doping can achieve n-type Mg_3Sb_2 with an electron concentration of 5×10^{20} cm^{-3} (Gorai et al. 2018), which was later verified experimentally in La-doped $Mg_3Sb_{2-x}Bi_x$ (Imasato et al. 2018).

The successful synthesis of n-type Mg_3Sb_2 contextualizes the importance of phase stability for tuning defect energetics. Often referred to as "phase boundary mapping", the technique charts the possible impurity phases in a material to gauge the thermodynamic equilibrium conditions and, specifically, the elemental chemical potentials (Borgsmiller et al. 2022). Since the chemical potentials ($\Delta\mu_i$, which is a part of μ_i in equation [2.15]) can be determined from a convex hull analysis, first-principles methods can be leveraged to rationally guide doping procedures of TE materials. For example, the analysis of native defects in Mg_3Sb_2 and the convex hull of the Mg–Sb binary system helped to realize n-type Mg_3Sb_2 under Mg-rich conditions (see Figure 2.6) (Ohno et al. 2018). In such thermodynamic conditions, the formation of acceptor-like Mg vacancies is suppressed, avoiding the possibility of electron compensation by the native defect. First-principles defect calculations have also revealed the high tunability of the carrier concentration (and, as a result, the electrical transport properties) in $Co_4Sn_6Te_6$ (Crawford et al. 2018), $Cu_2HgGeTe_4$ (Ortiz et al. 2019), Hg_2GeTe_4 (Qu et al. 2021), ZnSb (Wood et al. 2021), BiCuSeO (Toriyama et al. 2021), $CuInTe_2$ (Adamczyk et al. 2020) and $AgBiSe_2$ (Jang et al. 2022) through phase boundary mapping.

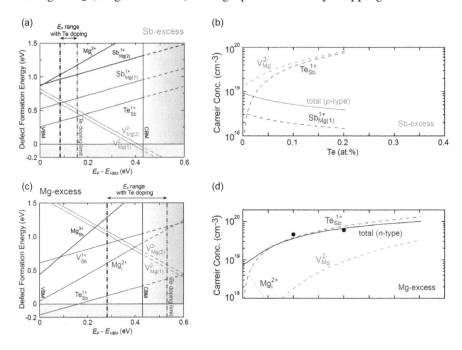

Figure 2.6. (a–b) The low formation energy of the acceptor-like Mg vacancy (V_{Mg}) in Mg_3Sb_2 under Sb-rich conditions limits the range of the Fermi energy (E_F) near the valence band. Mg_3Sb_2 is p-type under Mg-poor/Sb-excess conditions. (c–d) Under Mg-rich conditions, the achievable E_F range with Te doping is near the conduction band, since the formation of V_{Mg} is suppressed. As a result, n-type Mg_3Sb_2 is achieved through Te doping under Mg-rich conditions. The predicted electron concentrations are in agreement with experiments, which are denoted by black markers in (d). Reproduced from Ohno et al. (2018) with permission from Elsevier

Instead of focusing on the defect properties of a single material and finding ways to optimize its TE performance directly, recent studies have begun to explore the prospect of "inverse designing" materials that already possess the desired doping properties. One route that follows this philosophy is the chemical replacements in structure prototype (CRISP) method (Gorai et al. 2020). By generating both known and hypothetical compounds through chemical substitutions in a given prototype structure, the CRISP workflow aims to identify stable, high-performing and dopable TE materials within the generated chemical variants. The methodology was employed to search for yet-to-be-realized n-type TE materials within ABX Zintl phases, from which KSnSb was identified as a promising candidate (Gorai et al. 2020). Notably, KSnSb was found to be n-type-dopable, fulfilling the initial goal of finding an n-type Zintl candidate. Furthermore, by exploring the space of ABX_4 Zintl phases, three unreported phases ($NaAlSb_4$, $NaGaSb_4$, $CsInSb_4$) were predicted as high-performing n-type TE materials that are also n-type dopable (Qu et al. 2020).

In the effort to identify high-performing TE materials with desired doping properties, some intrinsic material properties have been shown to correlate well with the dopability of a material. For example, Zintl phases with average anion oxidation near -1 were found to be likely n-type-dopable due to the high formation energy of cation vacancies and the resulting absence of electron compensation (Gorai et al. 2019). Using this chemical guideline, the authors proposed that $ZnAs_2$, KSb and KBa_2As_5 are promising n-type TE candidates.

Despite the availability of high-performance computing resources, defect calculations are still laborious. Besides direct calculations of defect formation energetics, several alternative routes have been proposed, which can assess the dopability of materials in an indirect and more high-throughput manner. One such route is by predicting the absolute band edge positions using descriptors such as electron affinity and ionization potentials (Zhang et al. 1999; Walukiewicz 2001). Intuitively, n-type doping is challenging when the absolute energy of the conduction band minimum is high, and p-type doping is difficult when the absolute valence band maximum position is low. This guiding principle works well for many semiconductors, such as explaining the n-type dopability of the wide-gap ZnO and bipolar dopability of narrow-gap PbTe. In fact, the absolute band edge positions correlate with the dopabilities of rock-salt IV–VI compounds; while it is difficult to dope SnTe and GeTe n-types due to the high conduction band minimum energy, n-type PbTe is commonly achieved due to the lower conduction band minimum (Huo et al. 2021). Alternatively, the branch point energy was used to assess the dopabilities of ABXO (A,B = metals; X = S, Se, Te) phases (He et al. 2020). Since the branch point energy was found to be close to the middle of the band gap in BiAgXO compounds, He et al. (2020) concluded that the compounds are p-type-dopable, similar to the well-known p-type TE materials BiCuSeO (Zhao et al. 2014; Toriyama et al. 2021).

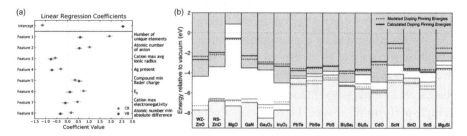

Figure 2.7. *(a) Linear regression model of the carrier concentration ranges of diamond-like semiconductors, showing how fundamental properties affect whether the material can be doped n-type (green) or p-type (purple). The coefficient value on the x-axis represents the weight of each feature in the linear regression model. (b) The calculated doping limit, quantified by the doping pinning energy, compared to the modeled limit as determined from features such as band edge positions and the average energies of atomic orbitals. Reproduced from Miller et al. (2018) and Goyal et al. (2020) with permission from Springer and the American Chemical Society*

A potential alternative approach to predicting the native defect energetics and dopability of a material is to leverage machine learning techniques. Miller et al. (2018) suggested an empirical model for dopability in diamond-like semiconductors (DLS), using a linear regression model developed from experimental carrier concentration data on 127 DLS compounds (see Figure 2.7(a)). Although accurate predictions of carrier concentrations in DLS are achieved within approximately one order of magnitude, the interpretation for the underlying physics behind the model is still unclear. Furthermore, the transferability of the model to other types of materials is uncertain. Goyal et al. (2020) developed a dopability model for ionic binary semiconductors based on intrinsic material descriptors. The model is validated against the pinning energies from first-principles calculations. As shown in Figure 2.7(b), both p-type and n-type dopabilities fit well for 16 classic binary semiconductors including III–Vs, II–VIs, group-III oxides, and lead and bismuth chalcogenides. It is argued that the dopability for semiconductors is a result of complex trade-offs between various intrinsic properties including electronic properties, nuclear repulsion and chemical potentials for constituent elements. This work points out that absolute band edge positions alone cannot be used to accurately describe the dopability of semiconductors. With the advent of diverse machine learning techniques, which will be described in the following section, there are many opportunities to employ such techniques towards understanding the thermodynamic and electronic properties of defects in semiconductors (Mannodi-Kanakkithodi et al. 2020, 2022; Mannodi-Kanakkithodi and Chan 2022; Polak et al. 2022).

2.2.3. Thermoelectric material search with high-throughput computations and machine learning

With the advances in computing power and methodologies, computational chemistry has made great strides in accelerating the discovery of functional materials with tailored properties. The ability to perform high-throughput (HT) ab initio calculations, in particular those based on density functional theory (DFT), has been instrumental in inorganic functional material discovery, including materials for TE applications. Such HT searches employ computationally tractable descriptors to assess the TE performance of large chemical spaces.

More recently, machine learning (ML) and material informatics approaches have been employed to further accelerate the search of TE materials. The rapid prediction of transport properties and TE performance with ML models has allowed the search to be expanded into even larger chemical spaces, including hypothetical compounds. Applications of material informations in TEs have made great strides in predicting relevant material properties such as lattice thermal conductivity and power factor, resulting in the prediction of many promising TE candidates, some of which have been synthesized and measured experimentally. There are a variety of ML techniques that have been used to predict TE properties, including decision trees and random forest regression, the sure independence screening and sparsifying operator method (Ouyang et al. 2018; Liu et al. 2020), gene expression programming (Abdellahi et al. 2015), neural networks (Laugier et al. 2018; Na et al. 2021; Zhu et al. 2021b), symbolic regression (Loftis et al. 2021), Bayesian optimization (Seko et al. 2015; Bassman et al. 2018) and active learning (Bassman et al. 2018; Hou et al. 2019; Sheng et al. 2020; Takagiwa et al. 2021; Tranås et al. 2022). These ML techniques are often trained on large databases of experimental and/or computational DFT data (Curtarolo et al. 2012; Gaultois et al. 2013; Jain et al. 2013; Gorai et al. 2016; Ricci et al. 2017; Choudhary et al. 2020a).

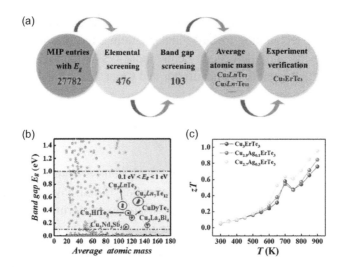

Figure 2.8. *(a) High-throughput workflow to screen for promising thermoelectric materials. (b) A map of ternary Cu-based pnictides and chalcogenides plotted as a function of two screening criteria – band gap and average atomic mass. (c) Experimental verification of the high performance of Cu_3ErTe_3-based thermoelectric materials. Reproduced from Wang et al. (2020a) with permission from Elsevier*

High-throughput computational searches employ descriptors that are identified a priori. For a detailed account of the descriptors and various HT searches for TE materials, we refer the reader to Gorai et al. (2017). Here, we discuss a few examples. Barreteau et al. (2019) searched for potential TE materials in the chemical space of *TMX* intermetallics, using thermodynamic stability and electronic structure features as screening criteria. One of the compounds identified in this study, half-Heusler TaFeSb, was later found to exhibit high TE performance experimentally, reaching a peak zT of 1.52 at 973 K when the Ta site is co-doped with V and Ti (Grytsiv et al. 2019; Zhu et al. 2019). Wang et al. (2020a) used band gap and average atomic mass to screen 27,782 ternary copper-based pnictides and chalcogenides (see Figure 2.8(a)), from which they identified Cu_3ErTe_3 as a promising TE candidate (see Figure 2.8(b)). The compound was subsequently synthesized experimentally, exhibiting a peak zT upwards of 1.0 when doped with Ag (see Figure 2.8(c)) (Wang et al. 2020a). Zhang et al. (2017) used various features of the crystal structure to screen Cu- and S-containing compounds for potential TE candidates. For example, one of the screening criteria was that Cu–S polyhedra in the structure must form a three-dimensional network, so that the carrier mobility is likely high. Using this and other screening criteria, the study identified and experimentally measured $Cu_6Fe_2S_8Sn$ and $Cu_{16}Fe_{4.3}S_{24}Sn_4Zn_{1.7}$ as high-performing TE materials (Zhang et al. 2017). Miller et al. (2017a) used a semi-empirical formulation of the TE quality factor to screen 735 oxide materials, from which they predicted SnO as a promising n-type candidate. However, the stability of SnO was found to be limited along the Sn–O binary phase space, leading to significant concentrations of SnO_2 during processing and low zT as a result (Miller et al. 2017a).

Often, regression techniques are used to identify relevant descriptors for properties of interest. While a slough of ML techniques have been applied to derive descriptors of transport properties, the models tend to recover similar physical descriptors for the property of interest. For example, ML models for predicting lattice thermal conductivity (κ_L) often identify the same material descriptors that appear in the analytical models of κ_L, such as the Slack and Debye–Callaway models. Random forest regression was used to identify heat capacity and phononic phase space volume as two prominent descriptors of κ_L (Carrete et al. 2014), which directly appear in the analytical κ_L models. A study of 120 binary, ternary and quaternary compounds found that κ_L depends strongly on physical quantities such as the maximum phonon frequency, integrated Grüneisen parameter, average atomic mass and the unit cell volume (Juneja et al. 2019). Bonding characteristics and electrical transport properties, such as the Seebeck coefficient and electrical conductivity, have also been used as descriptors to improve an ML model for κ_L (Juneja and Singh 2020). Correlations with the atomic masses of the constituent elements (Miyazaki et al. 2021), as well as the atomization enthalpy and unit cell volume (Wang et al. 2020b), have also been found using decision tree regression and extreme gradient boosting. For half-Heusler compounds, descriptors such as the lattice parameter, the atomic radii and atomic masses of the constituent elements are needed to predict κ_L (Miyazaki et al. 2021). For oxide TE materials, the ratio of oxygen to transition metal atoms was found to be a good descriptor of κ_L using a gradient-boosted tree model (Tewari et al. 2020).

Given that these techniques typically recover physically relevant descriptors that appear in analytical models of κ_L, it is interesting to compare the accuracies of the ML models to those of the analytical models. Chen et al. (2019) found that their ML model has accuracy comparable to the Slack and Debye–Callaway models. Juneja et al. (2019) found that predictions made by an ML model developed using Gaussian process regression (GPR) predicts κ_L

more accurately than the Slack model by an order of magnitude. Symbolic regression has also been used to develop models of κ_L, which outperform the Slack model (Loftis et al. 2021).

Material descriptors that are relevant for electrical transport properties have also been elucidated by regression techniques. Gradient-boosted decision trees identified that polarizability, atomic masses and radii, and refractive index are necessary features to quantitatively predict the Seebeck coefficient and power factor of materials (Choudhary et al. 2020b). In addition to elemental descriptors, predictions of the Seebeck coefficient and electrical conductivity have been found to improve with chemical bonding-based descriptors (Juneja and Singh 2020).

ML models are now routinely employed to rapidly identify promising TE material candidates from large chemical spaces. Many studies have trained ML models to predict low-κ_L materials for potential TE applications. Carrete et al. (2014) trained an ML model on a large dataset of half-Heusler compounds, predicting that PtLaSb, RhLaTe and SbNaSr should exhibit low κ_L. The low κ_L of PtLaSb was later confirmed by solving the Peierls–Boltzmann transport equation and through ab initio molecular dynamics (Feng et al. 2020), inspiring further investigations on its TE properties (Xue et al. 2016). It is interesting to note, however, that SbNaSr, another half-Heusler compound that was predicted to have a low κ_L (Carrete et al. 2014), was later found to be dynamically unstable by first-principles calculations (Feng et al. 2020), suggesting that ML-based predictions of κ_L should be supported by more in-depth first-principles calculations. Seko et al. (2015) used Bayesian optimization techniques to screen 221 low-κ_L compounds out of a dataset of 54,779 compounds considered, using only the volume, density and elemental properties as descriptors. One of the compounds predicted to have low κ_L, $Cs_2PdCl_4I_2$, was investigated further using first-principles calculations (Guo 2016; Li and Yang 2016), predicting a thermal conductivity of 0.31 W/mK with maximum zT of 0.7 (Guo 2016). Wang et al. (2020b) used extreme gradient boosting to predict low-κ_L compounds, which resulted in the prediction of $BiTe_2Tl$, $Br_2Cs_2F_2$, Au_3CsSe_2 and Cl_2CsI as low-κ_L compounds. Zhang et al. used a web-based recommendation engine (Gaultois et al. 2016) to predict low-κ_L compounds in the $RE_4M_2XGe_4$ chemical space. Although intermetallic compounds typically display high κ_L on the order of $10^1 - 10^2$ W/mK (Terada et al. 2002), $Nd_4Mn_2InGe_4$ and $Nd_4Mn_2AgGe_4$ were predicted, and subsequently shown experimentally, to possess $\kappa_L < 10$ W/mK. Zhu et al. (2021b) revealed using random forest regression and crystal graph convolutional neural networks that rare-earth chalcogenides rank in the lowest 5% of the predicted κ_L, which was experimentally verified for several compounds in this family (see Figure 2.9). Interestingly, the study found by performing additional electrical transport measurements on Bi-doped Er_2Te_3 and Y_2Te_3 that the zT reaches upwards of 1, suggesting that rare-earth chalcogenides are promising TE candidates (Zhu et al. 2021b; Toriyama et al. 2022c).

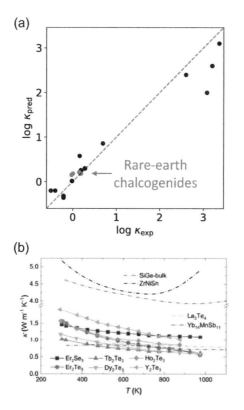

Figure 2.9. *(a) Lattice thermal conductivity (κ_L) predicted by a deep learning model (uses crystal graph convolutional neural network) that is trained on DFT and experimental data. The predicted κ_L are in fairly good agreement with measured κ_L. Rare-earth chalcogenides are among the materials with the lowest κ_L. (b) Experimental validation of the prediction that rare-earth chalcogenides have low κ_L, comparable to other well-known thermoelectric materials such as $Yb_{14}MnSb_{11}$. Data in panel (a) is adapted from Zhu et al. (2021b). Panel (b) is reproduced from Zhu et al. (2021b) with permission from the Royal Society of Chemistry*

ML models have also been used to improve electrical transport properties of TE materials. For example, the power factor of n-type $Al_2Fe_3Si_3$ was enhanced by $\sim$40% by continuously feeding experimentally measured data to a developing ML model (see Figure 2.10(a)) (Hou et al. 2019). In this study, Hou et al. began with an initial ML model that was constructed from measured TE property data of $Al_{23.5+x}Fe_{36.5}Si_{40-x}$ at several compositions, which was used to predict new compositions with enhanced power factors. The predictions were experimentally synthesized, and the measured data on the new batch of samples was fed into the ML model to make new predictions. An overall improvement in the maximum power factor from 525 μW/mK^2 at $x = 0$ to 670 μW/mK^2 at $x = 0.9$ was achieved through this method (see Figure 2.10(b)) (Hou et al. 2019). A similar iterative approach was used to improve the power factor of Co-doped Al–Fe–Si-based TE materials (Takagiwa et al. 2021). Iwasaki et al. (2019a, 2019b) used ML models to study the anomalous Nernst effect in spin-driven TE systems. They found a positive correlation between the orbital angular momentum and the spin-driven thermopower when studying a rare-earth-substituted yttrium iron garnet system, from which they predicted and verified experimentally that $Fe_{0.7}Pt_{0.3}Sm_{0.05}$ has a higher thermopower than FePt alloys (Iwasaki et al. 2019b). Afterwards, Iwasaki et al. (2019a) studied the $M_{100-x}Pt_x$ (M = Fe, Co and Ni) system using similar ML approaches. In addition to recovering the known physics of spin-driven TE materials (e.g. that the anomalous Nernst effect arises only in magnetic materials), they found that the effect can be enhanced by the local spin polarization and orbital moment of Pt. Using this knowledge, Iwasaki et al. (2019a) predicted that $Co_{50}Pt_{50}N_x$ will exhibit enhanced spin-driven thermopower compared to the parent CoPt alloy, which was confirmed experimentally.

ML models that simultaneously predict the electrical and thermal transport properties (and, as a result, zT) have also been developed. Gaultois et al. (2016) built a web-based recommendation engine that simultaneously optimizes the Seebeck coefficient, electrical conductivity and thermal conductivity for a given chemical composition. The engine predicted that $RE_{12}Co_5Bi$ (RE = Gd, Er) displays properties that are optimal for thermoelectricity, as verified in the same study experimentally (Gaultois et al. 2016). While the recommendation engine correctly predicted the thermal conductivity at room temperature, experimental validation revealed that the thermal diffusivity of $RE_{12}Co_5Bi$ increases with temperature up to 800 K, demonstrating that the experimental verification of ML-predicted compounds may lead to simultaneously revealing anomalous behaviors in materials. Moreover, attempts to synthesize $Gd_{12}Co_5Bi$ led to the discovery of cubic $Gd_{12}Co_{5.3}Bi$ with lower electrical resistivity and thermal conductivity (Oliynyk et al. 2016), demonstrating that ML-based predictions of TE candidates may lead to experimental realizations of similar compounds with comparable, or even better, performance. Bassman et al. (2018) used an active learning method and a Gaussian process regression model to develop stacked heterostructures for TE applications. By developing a model to predict the electronic fitness function (Xing et al. 2017) for doped heterostructures, they predicted that the three-layer $MoSe_2$–WS_2–WS_2 and WSe_2–WTe_2–WSe_2 system is the best n-type candidate, and WTe_2–$MoTe_2$–WTe_2 and $MoSe_2$–WSe_2–WSe_2 are the best p-type candidates.

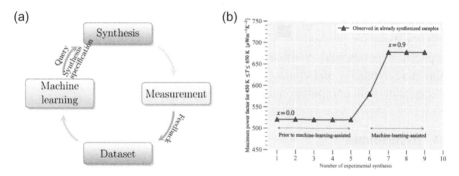

Figure 2.10. *(a) A machine learning-assisted workflow for achieving high performance in thermoelectric materials. (b) Demonstration of the workflow for improving the power factor of $Al_{23.5+x}Fe_{36.5}Si_{40-x}$. The power factor is increased significantly from 520 $\mu W/mK^2$ at $x = 0$ to 670 $\mu W/mK^2$ at $x = 0.9$. Reproduced from Hou et al. (2019) with permission from the American Chemical Society*

2.3. Outlook

The discovery and design of thermoelectric materials have greatly benefited from the advances in computational methods and development of software packages. In this chapter, we have mapped the historical contributions of computations and data-driven approaches in thermoelectric material research and highlighted some recent advances. While this book chapter solely focuses on first-principles computations and machine learning in TE material development, we must also acknowledge the important role of micro-scale continuum modeling and specifically device modeling.

There are outstanding challenges in the computational modeling of thermoelectric materials that need to be addressed. There is a need to develop computationally efficient and more accurate methods for calculating scattering rates, and incorporating temperature-dependent effects and disorder in the calculation of transport and thermoelectric properties. Computational TE material discovery, by and large, has focused on stoichiometric and ordered structures that are documented in crystallographic databases, but a massively larger search space of hypothetical compounds, including disordered phases and alloys, remains to be explored. Computational methods need to be developed to assess large chemical spaces that contain one to two orders of magnitude more structures compared to traditional high-throughput DFT searches. Data-driven methods will likely play a key role in this development. Finally, it is imperative that computational findings are experimentally verified to avoid overwhelming the thermoelectric literature with unreasonable predictions. Currently, the rate of computational predictions outpaces experimental verification. The development of high-throughput solid-state synthesis and characterization will allow experimental verification to keep pace with computational predictions.

2.4. References

Abdellahi, M., Bahmanpour, M., Bahmanpour, M. (2015). Modeling Seebeck coefficient of $Ca_{3-x}M_xCo_4O_9$ (M= Sr, Pr, Ga, Ca, Ba, La, Ag) thermoelectric ceramics. *Ceram. Int.*, 41(1), 345.

Adamczyk, J.M., Gomes, L.C., Qu, J., Rome, G.A., Baumann, S.M., Ertekin, E., Toberer, E.S. (2020). Native defect engineering in $CuInTe_2$. *Chem. Mater.*, 33(1), 359.

Anand, S., Wood, M., Xia, Y., Wolverton, C., Snyder, G.J. (2019). Double half-Heuslers. *Joule*, 3(5), 1226.

Anand, S., Toriyama, M.Y., Wolverton, C., Haile, S.M., Snyder, G.J. (2022). A convergent understanding of charged defects. *Acc. Mater. Res.*, 3(7), 685.

Arrigoni, M. and Madsen, G.K. (2021). Spinney: Post-processing of first-principles calculations of point defects in semiconductors with Python. *Comput. Phys. Commun.*, 264, 107946.

Ashcroft, N.W. and Mermin, N.D. (1976). *Solid State Physics*, vol. 3. Cengage Learning, Boston, MA.

Askerov, B.M. (1994). *Electron Transport Phenomena in Semiconductors*. World Scientific, Singapore.

Aydemir, U., Pöhls, J.-H., Zhu, H., Hautier, G., Bajaj, S., Gibbs, Z.M., Chen, W., Li, G., Ohno, S., Broberg, D. et al. (2016). YCuTe$_2$: A member of a new class of thermoelectric materials with CuTe$_4$-based layered structure. *J. Mater. Chem. A*, 4(7), 2461.

Bai, Q., Zhang, X., Shan, B., Shi, X., Sun, C., Lin, S., Li, W., Pei, Y. (2021). Thermoelectric transport properties of TmAg$_x$Cu$_{1-x}$Te$_2$ solid solutions. *J. Materiomics*, 7(4), 886.

Baroni, S., De Gironcoli, S., Dal Corso, A., Giannozzi, P. (2001). Phonons and related crystal properties from density-functional perturbation theory. *Rev. Mod. Phys.*, 73(2), 515.

Barreteau, C., Crivello, J.-C., Joubert, J.-M., Alleno, E. (2019). Looking for new thermoelectric materials among TMX intermetallics using high-throughput calculations. *Comp. Mater. Sci.*, 156, 96.

Bassman, L., Rajak, P., Kalia, R.K., Nakano, A., Sha, F., Sun, J., Singh, D.J., Aykol, M., Huck, P., Persson, K. et al. (2018). Active learning for accelerated design of layered materials. *npj Comput. Mater.*, 4(1), 1.

Bernardi, M. (2016). First-principles dynamics of electrons and phonons. *Eur. Phys. J. B*, 89(11), 1.

Bilc, D.I., Hautier, G., Waroquiers, D., Rignanese, G.-M., Ghosez, P. (2015). Low-dimensional transport and large thermoelectric power factors in bulk semiconductors by band engineering of highly directional electronic states. *Phys. Rev. Lett.*, 114(13), 136601.

Bjerg, L., Madsen, G.K., Iversen, B.B. (2011). Enhanced thermoelectric properties in zinc antimonides. *Chem. Mater.*, 23(17), 3907.

Bjerg, L., Iversen, B.B., Madsen, G.K. (2014). Modeling the thermal conductivities of the zinc antimonides ZnSb and Zn$_4$Sb$_3$. *Phys. Rev. B*, 89(2), 024304.

de Boor, J. (2021). On the applicability of the single parabolic band model to advanced thermoelectric materials with complex band structures. *J. Materiomics*, 7(3), 603.

Borgsmiller, L., Zavanelli, D., Snyder, G.J. (2022). Phase-boundary mapping to engineer defects in thermoelectric materials. *PRX Energy*, 1(2), 022001.

Bottin, F., Bieder, J., Bouchet, J. (2020). a-TDEP: Temperature dependent effective potential for ABINIT – Lattice dynamic properties including anharmonicity. *Comput. Phys. Commun.*, 254, 107301.

Broberg, D., Medasani, B., Zimmermann, N.E., Yu, G., Canning, A., Haranczyk, M., Asta, M., Hautier, G. (2018). PyCDT: A Python toolkit for modeling point defects in semiconductors and insulators. *Comput. Phys. Commun.*, 226, 165.

Brod, M.K. and Snyder, G.J. (2021). Orbital chemistry of high valence band convergence and low-dimensional topology in PbTe. *J. Mater. Chem. A*, 9(20), 12119.

Burstein, E. (1954). Anomalous optical absorption limit in InSb. *Phys. Rev.*, 93(3), 632.

Cao, J., Querales-Flores, J.D., Murphy, A.R., Fahy, S., Savić, I. (2018). Dominant electron–phonon scattering mechanisms in n-type PbTe from first principles. *Phys. Rev. B*, 98(20), 205202.

Carrete, J., Li, W., Mingo, N., Wang, S., Curtarolo, S. (2014). Finding unprecedentedly low-thermal-conductivity half-Heusler semiconductors via high-throughput materials modeling. *Phys. Rev. X*, 4(1), 011019.

Carrete, J., Vermeersch, B., Katre, A., van Roekeghem, A., Wang, T., Madsen, G.K., Mingo, N. (2017). almaBTE: A solver of the space–time dependent Boltzmann transport equation for phonons in structured materials. *Comput. Phys. Commun.*, 220, 351.

Chen, X., Parker, D., Singh, D.J. (2013). Importance of non-parabolic band effects in the thermoelectric properties of semiconductors. *Sci. Rep.*, 3(1), 1.

Chen, W., Pöhls, J.-H., Hautier, G., Broberg, D., Bajaj, S., Aydemir, U., Gibbs, Z.M., Zhu, H., Asta, M., Snyder, G.J. et al. (2016). Understanding thermoelectric properties from high-throughput calculations: Trends, insights, and comparisons with experiment. *J. Mater. Chem. C*, 4(20), 4414.

Chen, L., Tran, H., Batra, R., Kim, C., Ramprasad, R. (2019). Machine learning models for the lattice thermal conductivity prediction of inorganic materials. *Comp. Mater. Sci.*, 170, 109155.

Chernatynskiy, A. and Phillpot, S.R. (2015). Phonon transport simulator (PhonTS). *Comput. Phys. Commun.*, 192, 196.

Choudhary, K., Garrity, K.F., Reid, A.C., DeCost, B., Biacchi, A.J., Walker, A.R.H., Trautt, Z., Hattrick-Simpers, J., Kusne, A.G., Centrone, A. et al. (2020a). The joint automated repository for various integrated simulations (JARVIS) for data-driven materials design. *npj Comput. Mater.*, 6(1), 1.

Choudhary, K., Garrity, K.F., Tavazza, F. (2020b). Data-driven discovery of 3D and 2D thermoelectric materials. *J. Phys. Condens. Mat.*, 32(47), 475501.

Crawford, C.M., Ortiz, B.R., Gorai, P., Stevanovic, V., Toberer, E.S. (2018). Experimental and computational phase boundary mapping of $Co_4Sn_6Te_6$. *J. Mater. Chem. A*, 6(47), 24175.

Curtarolo, S., Setyawan, W., Wang, S., Xue, J., Yang, K., Taylor, R.H., Nelson, L.J., Hart, G.L., Sanvito, S., Buongiorno-Nardelli, M. et al. (2012). AFLOWLIB.ORG: A distributed materials properties repository from high-throughput ab initio calculations. *Comp. Mater. Sci.*, 58, 227.

Deng, T., Wu, G., Sullivan, M.B., Wong, Z.M., Hippalgaonkar, K., Wang, J.-S., Yang, S.-W. (2020). EPIC STAR: A reliable and efficient approach for phonon- and impurity-limited charge transport calculations. *npj Comput. Mater.*, 6(1), 1.

Ding, J., Liu, C., Xi, L., Xi, J., Yang, J. (2021). Thermoelectric transport properties in chalcogenides ZnX (X= S, Se): From the role of electron-phonon couplings. *J. Materiomics*, 7(2), 310.

Drude, P. (1900). Zur elektronentheorie der metalle. *Ann. Phys.*, 306(3), 566.

D'Souza, R., Cao, J., Querales-Flores, J.D., Fahy, S., Savić, I. (2020). Electron–phonon scattering and thermoelectric transport in p-type PbTe from first principles. *Phys. Rev. B*, 102(11), 115204.

Du, B., Zhang, R., Chen, K., Mahajan, A., Reece, M.J. (2017). The impact of lone-pair electrons on the lattice thermal conductivity of the thermoelectric compound $CuSbS_2$. *J. Mater. Chem. A*, 5(7), 3249.

Dylla, M.T., Kang, S.D., Snyder, G.J. (2019). Effect of two-dimensional crystal orbitals on fermi surfaces and electron transport in three-dimensional perovskite oxides. *Angew. Chem.*, 131(17), 5557.

Faghaninia, A., Ager III, J.W., Lo, C.S. (2015). Ab initio electronic transport model with explicit solution to the linearized Boltzmann transport equation. *Phys. Rev. B*, 91(23), 235123.

Faghaninia, A., Yu, G., Aydemir, U., Wood, M., Chen, W., Rignanese, G.-M., Snyder, G.J., Hautier, G., Jain, A. (2017). A computational assessment of the electronic, thermoelectric, and defect properties of bournonite ($CuPbSbS_3$) and related substitutions. *Phys. Chem. Chem. Phys.*, 19(9), 6743.

Fan, T. and Oganov, A.R. (2020). AICON: A program for calculating thermal conductivity quickly and accurately. *Comput. Phys. Commun.*, 251, 107074.

Fan, T. and Oganov, A.R. (2021). AICON2: A program for calculating transport properties quickly and accurately. *Comput. Phys. Commun.*, 266, 108027.

Fang, T., Zheng, S., Chen, H., Cheng, H., Wang, L., Zhang, P. (2016). Electronic structure and thermoelectric properties of p-type half-Heusler compound NbFeSb: A first-principles study. *RSC Adv.*, 6(13), 10507.

Feng, T., Lindsay, L., Ruan, X. (2017). Four-phonon scattering significantly reduces intrinsic thermal conductivity of solids. *Phys. Rev. B*, 96(16), 161201.

Feng, Z., Fu, Y., Putatunda, A., Zhang, Y., Singh, D.J. (2019). Electronic structure as a guide in screening for potential thermoelectrics: Demonstration for half-Heusler compounds. *Phys. Rev. B*, 100(8), 085202.

Feng, Z., Fu, Y., Zhang, Y., Singh, D.J. (2020). Characterization of rattling in relation to thermal conductivity: Ordered half-Heusler semiconductors. *Phys. Rev. B*, 101(6), 064301.

Franz, R. and Wiedemann, G. (1853). Ueber die wärme-leitungsfähigkeit der metalle. *Ann. Phys.*, 165(8), 497.

Freysoldt, C., Grabowski, B., Hickel, T., Neugebauer, J., Kresse, G., Janotti, A., Van de Walle, C.G. (2014). First-principles calculations for point defects in solids. *Rev. Mod. Phys.*, 86(1), 253.

Ganose, A.M., Park, J., Faghaninia, A., Woods-Robinson, R., Persson, K.A., Jain, A. (2021). Efficient calculation of carrier scattering rates from first principles. *Nat. Commun.*, 12(1), 1.

Garrity, K.F. (2016). First-principles search for n-type oxide, nitride, and sulfide thermoelectrics. *Phys. Rev. B*, 94(4), 045122.

Gaultois, M.W., Sparks, T.D., Borg, C.K., Seshadri, R., Bonificio, W.D., Clarke, D.R. (2013). Data-driven review of thermoelectric materials: Performance and resource considerations. *Chem. Mater.*, 25(15), 2911.

Gaultois, M.W., Oliynyk, A.O., Mar, A., Sparks, T.D., Mulholland, G.J., Meredig, B. (2016). Perspective: Web-based machine learning models for real-time screening of thermoelectric materials properties. *APL Mater.*, 4(5), 053213.

Giustino, F. (2017). Electron–phonon interactions from first principles. *Rev. Mod. Phys.*, 89(1), 015003.

Giustino, F., Cohen, M.L., Louie, S.G. (2007). Electron–phonon interaction using Wannier functions. *Phys. Rev. B*, 76(16), 165108.

Goldsmid, H. (2013). *Thermoelectric Refrigeration.* Springer, New York.

Goldsmid, J. (2017). *The Physics of Thermoelectric Energy Conversion.* Morgan & Claypool Publishers, San Rafael, CA.

Gorai, P., Gao, D., Ortiz, B., Miller, S., Barnett, S.A., Mason, T., Lv, Q., Stevanović, V., Toberer, E.S. (2016). TE Design Lab: A virtual laboratory for thermoelectric material design. *Comp. Mater. Sci.*, 112, 368.

Gorai, P., Stevanović, V., Toberer, E.S. (2017). Computationally guided discovery of thermoelectric materials. *Nat. Rev. Mater.*, 2(9), 1.

Gorai, P., Ortiz, B.R., Toberer, E.S., Stevanović, V. (2018). Investigation of n-type doping strategies for Mg_3Sb_2. *J. Mater. Chem. A*, 6(28), 13806.

Gorai, P., Goyal, A., Toberer, E.S., Stevanović, V. (2019). A simple chemical guide for finding novel n-type dopable Zintl pnictide thermoelectric materials. *J. Mater. Chem. A*, 7(33), 19385.

Gorai, P., Ganose, A., Faghaninia, A., Jain, A., Stevanović, V. (2020). Computational discovery of promising new n-type dopable ABX Zintl thermoelectric materials. *Mater. Horiz.*, 7(7), 1809.

Goyal, A., Gorai, P., Peng, H., Lany, S., Stevanović, V. (2017a). A computational framework for automation of point defect calculations. *Comp. Mater. Sci.*, 130, 1.

Goyal, A., Gorai, P., Toberer, E.S., Stevanović, V. (2017b). First-principles calculation of intrinsic defect chemistry and self-doping in PbTe. *npj Comput. Mater.*, 3(1), 1.

Goyal, A., Gorai, P., Anand, S., Toberer, E.S., Snyder, G.J., Stevanovic, V. (2020). On the dopability of semiconductors and governing material properties. *Chem. Mater.*, 32(11), 4467.

Graziosi, P., Kumarasinghe, C., Neophytou, N. (2019). Impact of the scattering physics on the power factor of complex thermoelectric materials. *J. Appl. Phys.*, 126(15), 155701.

Grytsiv, A., Romaka, V., Watson, N., Rogl, G., Michor, H., Hinterleitner, B., Puchegger, S., Bauer, E., Rogl, P. (2019). Thermoelectric half-Heusler compounds TaFeSb and $Ta_{1-x}Ti_xFeSb$ ($0 \leq x \leq 0.11$): Formation and physical properties. *Intermetallics*, 111, 106468.

Gunatilleke, W.D., Juneja, R., Ojo, O.P., May, A.F., Wang, H., Lindsay, L., Nolas, G.S. (2021). Intrinsic anharmonicity and thermal properties of ultralow thermal conductivity $Ba_6Sn_6Se_{13}$. *Phys. Rev. Mater.*, 5(8), 085002.

Guo, S.-D. (2016). Potential thermoelectric material $Cs_2[PdCl_4]I_2$: A first-principles study. *Mater. Res. Express*, 3(8), 085903.

Han, Z., Yang, X., Li, W., Feng, T., Ruan, X. (2022). FourPhonon: An extension module to ShengBTE for computing four-phonon scattering rates and thermal conductivity. *Comput. Phys. Commun.*, 270, 108179.

He, J., Amsler, M., Xia, Y., Naghavi, S.S., Hegde, V.I., Hao, S., Goedecker, S., Ozoliņš, V., Wolverton, C. (2016). Ultralow thermal conductivity in full Heusler semiconductors. *Phys. Rev. Lett.*, 117(4), 046602.

He, J., Hao, S., Xia, Y., Naghavi, S.S., Ozolins, V., Wolverton, C. (2017). Bi_2PdO_4: A promising thermoelectric oxide with high power factor and low lattice thermal conductivity. *Chem. Mater.*, 29(6), 2529.

He, J., Yao, Z., Hegde, V.I., Naghavi, S.S., Shen, J., Bushick, K.M., Wolverton, C. (2020). Computational discovery of stable heteroanionic oxychalcogenides ABXO (A, B = metals; X = S, Se, and Te) and their potential applications. *Chem. Mater.*, 32(19), 8229.

Hellman, O. and Abrikosov, I.A. (2013). Temperature-dependent effective third-order interatomic force constants from first principles. *Phys. Rev. B*, 88(14), 144301.

Hellman, O., Abrikosov, I.A., Simak, S. (2011). Lattice dynamics of anharmonic solids from first principles. *Phys. Rev. B*, 84(18), 180301.

Hellman, O., Steneteg, P., Abrikosov, I.A., Simak, S.I. (2013). Temperature dependent effective potential method for accurate free energy calculations of solids. *Phys. Rev. B*, 87(10), 104111.

Hicks, L.D. and Dresselhaus, M.S. (1993). Effect of quantum-well structures on the thermoelectric figure of merit. *Phys. Rev. B*, 47(19), 12727.

Hou, Z., Takagiwa, Y., Shinohara, Y., Xu, Y., Tsuda, K. (2019). Machine-learning-assisted development and theoretical consideration for the $Al_2Fe_3Si_3$ thermoelectric material. *ACS Appl. Mater. Interfaces*, 11(12), 11545.

Huang, M., Zheng, Z., Dai, Z., Guo, X., Wang, S., Jiang, L., Wei, J., Chen, S. (2022). DASP: Defect and dopant ab-initio simulation package. *J. Semicond.*, 43(4), 042101.

Huo, H., Wang, Y., Xi, L., Yang, J., Zhang, W. (2021). The variation of intrinsic defects in XTe (X = Ge, Sn, and Pb) induced by the energy positions of valence band maxima. *J. Mater. Chem. C*, 9(17), 5765.

Imasato, K., Wood, M., Kuo, J.J., Snyder, G.J. (2018). Improved stability and high thermoelectric performance through cation site doping in n-type La-doped $Mg_3Sb_{1.5}Bi_{0.5}$. *J. Mater. Chem. A*, 6(41), 19941.

Iwasaki, Y., Sawada, R., Stanev, V., Ishida, M., Kirihara, A., Omori, Y., Someya, H., Takeuchi, I., Saitoh, E., Yorozu, S. (2019a). Identification of advanced spin-driven thermoelectric materials via interpretable machine learning. *npj Comput. Mater.*, 5(1), 1.

Iwasaki, Y., Takeuchi, I., Stanev, V., Kusne, A.G., Ishida, M., Kirihara, A., Ihara, K., Sawada, R., Terashima, K., Someya, H. et al. (2019b). Machine-learning guided discovery of a new thermoelectric material. *Sci. Rep.*, 9(1), 1.

Jain, A., Ong, S.P., Hautier, G., Chen, W., Richards, W.D., Dacek, S., Cholia, S., Gunter, D., Skinner, D., Ceder, G. et al. (2013). Commentary: The Materials Project: A materials genome approach to accelerating materials innovation. *APL Mater.*, 1(1), 011002.

Jang, H., Toriyama, M.Y., Abbey, S., Frimpong, B., Male, J.P., Snyder, G.J., Jung, Y.S., Oh, M.-W. (2022). Suppressing charged cation antisites via Se vapor annealing enables p-type dopability in $AgBiSe_2$–SnSe thermoelectrics. *Adv. Mater.*, 34(38), 2204132.

Juneja, R. and Singh, A.K. (2020). Unraveling the role of bonding chemistry in connecting electronic and thermal transport by machine learning. *J. Mater. Chem. A*, 8(17), 8716.

Juneja, R., Yumnam, G., Satsangi, S., Singh, A.K. (2019). Coupling the high-throughput property map to machine learning for predicting lattice thermal conductivity. *Chem. Mater.*, 31(14), 5145.

Jung, Y.-K., Han, I.T., Kim, Y.C., Walsh, A. (2021). Prediction of high thermoelectric performance in the low-dimensional metal halide $Cs_3Cu_2I_5$. *npj Comput. Mater.*, 7(1), 1.

Kang, S.D., Dylla, M., Snyder, G.J. (2018). Thermopower-conductivity relation for distinguishing transport mechanisms: Polaron hopping in CeO_2 and band conduction in $SrTiO_3$. *Phys. Rev. B*, 97(23), 235201.

Katsura, Y., Kumagai, M., Kodani, T., Kaneshige, M., Ando, Y., Gunji, S., Imai, Y., Ouchi, H., Tobita, K., Kimura, K. et al. (2019). Data-driven analysis of electron relaxation times in PbTe-type thermoelectric materials. *Sci. Technol. Adv. Mat.*, 20(1), 511.

Kayser, P., Serrano-Sanchez, F., Dura, O.J., Fauth, F., Alonso, J.A. (2020). Experimental corroboration of the thermoelectric performance of Bi_2PdO_4 oxide and Pb-doped derivatives. *J. Mater. Chem. C*, 8(16), 5509.

Kittel, C. and McEuen, P. (1996). *Introduction to Solid State Physics*, vol. 8. Wiley, New York.

Koumoto, K. and Mori, T. (2013). *Thermoelectric Nanomaterials*. Springer, Heidelberg.

Kumar, N. and Bera, C. (2021). Theoretical prediction of thermoelectric properties of n-type binary Zintl compounds (KSb and KBi). *Physica B*, 619, 413206.

Lai, W., Wang, Y., Morelli, D.T., Lu, X. (2015). From bonding asymmetry to anharmonic rattling in $Cu_{12}Sb_4S_{13}$ tetrahedrites: When lone-pair electrons are not so lonely. *Adv. Funct. Mater.*, 25(24), 3648.

Lany, S. and Zunger, A. (2008). Assessment of correction methods for the band-gap problem and for finite-size effects in supercell defect calculations: Case studies for ZnO and GaAs. *Phys. Rev. B*, 78, 235104.

Laugier, L., Bash, D., Recatala, J., Ng, H.K., Ramasamy, S., Foo, C.-S., Chandrasekhar, V.R., Hippalgaonkar, K. (2018). Predicting thermoelectric properties from crystal graphs and material descriptors-first application for functional materials. *arXiv preprint*. arXiv:1811.06219.

Lee, S., Esfarjani, K., Luo, T., Zhou, J., Tian, Z., Chen, G. (2014a). Resonant bonding leads to low lattice thermal conductivity. *Nat. Commun.*, 5(1), 1.

Lee, Y., Lo, S.-H., Chen, C., Sun, H., Chung, D.-Y., Chasapis, T.C., Uher, C., Dravid, V.P., Kanatzidis, M.G. (2014b). Contrasting role of antimony and bismuth dopants on the thermoelectric performance of lead selenide. *Nat. Commun.*, 5(1), 1.

Li, W. and Yang, G. (2016). Structure and thermoelectric properties of the quaternary compound $Cs_2[PdCl_4]I_2$ with ultralow lattice thermal conductivity. *Europhys. Lett.*, 113(5), 57007.

Li, W., Carrete, J., Katcho, N.A., Mingo, N. (2014). ShengBTE: A solver of the Boltzmann transport equation for phonons. *Comput. Phys. Commun.*, 185(6), 1747.

Li, R., Li, X., Xi, L., Yang, J., Singh, D.J., Zhang, W. (2019). High-throughput screening for advanced thermoelectric materials: Diamond-like ABX_2 compounds. *ACS Appl. Mater. Interfaces*, 11(28), 24859.

Li, X., Zhang, Z., Xi, J., Singh, D.J., Sheng, Y., Yang, J., Zhang, W. (2021a). TransOpt. A code to solve electrical transport properties of semiconductors in constant electron–phonon coupling approximation. *Comp. Mater. Sci.*, 186, 110074.

Li, Z., Graziosi, P., Neophytou, N. (2021b). Deformation potential extraction and computationally efficient mobility calculations in silicon from first principles. *Phys. Rev. B*, 104(19), 195201.

Lin, H., Tan, G., Shen, J.-N., Hao, S., Wu, L.-M., Calta, N., Malliakas, C., Wang, S., Uher, C., Wolverton, C. et al. (2016). Concerted rattling in $CsAg_5Te_3$ leading to ultralow thermal conductivity and high thermoelectric performance. *Angew. Chem.*, 128(38), 11603.

Lin, H., Chen, H., Ma, N., Zheng, Y.-J., Shen, J.-N., Yu, J.-S., Wu, X.-T., Wu, L.-M. (2017). Syntheses, structures, and thermoelectric properties of ternary tellurides: $RECuTe_2$ (RE = Tb–Er). *Inorg. Chem. Front.*, 4(8), 1273.

Lindsay, L., Katre, A., Cepellotti, A., Mingo, N. (2019). Perspective on ab initio phonon thermal transport. *J. Appl. Phys.*, 126(5), 050902.

Liu, Z., Geng, H., Shuai, J., Wang, Z., Mao, J., Wang, D., Jie, Q., Cai, W., Sui, J., Ren, Z. (2015). The effect of nickel doping on electron and phonon transport in the n-type nanostructured thermoelectric material CoSbS. *J. Mater. Chem. C*, 3(40), 10442.

Liu, T.-H., Zhou, J., Liao, B., Singh, D.J., Chen, G. (2017). First-principles mode-by-mode analysis for electron-phonon scattering channels and mean free path spectra in GaAs. *Phys. Rev. B*, 95(7), 075206.

Liu, Z., Guo, S., Wu, Y., Mao, J., Zhu, Q., Zhu, H., Pei, Y., Sui, J., Zhang, Y., Ren, Z. (2019). Design of high-performance disordered half-Heusler thermoelectric materials using 18-electron rule. *Adv. Funct. Mater.*, 29(44), 1905044.

Liu, J., Han, S., Cao, G., Zhou, Z., Sheng, C., Liu, H. (2020). A high-throughput descriptor for prediction of lattice thermal conductivity of half-Heusler compounds. *J. Phys. D Appl. Phys.*, 53(31), 315301.

Loftis, C., Yuan, K., Zhao, Y., Hu, M., Hu, J. (2021). Lattice thermal conductivity prediction using symbolic regression and machine learning. *J. Phys. Chem. A*, 125(1), 435.

Lundstrom, M. (2002). *Fundamentals of Carrier Transport*. Cambridge University Press, Cambridge.

Ma, J., Chen, Y., Li, W. (2018). Intrinsic phonon-limited charge carrier mobilities in thermoelectric SnSe. *Phys. Rev. B*, 97(20), 205207.

Ma, J., Nissimagoudar, A.S., Wang, S., Li, W. (2020). High thermoelectric figure of merit of full-Heusler Ba_2AuX (X = As, Sb, and Bi). *Phys. Status Solidi*, 14(6), 2000084.

Madsen, G.K. (2006). Automated search for new thermoelectric materials: The case of LiZnSb. *J. Am. Chem. Soc.*, 128(37), 12140.

Madsen, G.K. and Singh, D.J. (2006). BoltzTraP. A code for calculating band-structure dependent quantities. *Comput. Phys. Commun.*, 175(1), 67.

Madsen, G.K., Carrete, J., Verstraete, M.J. (2018). BoltzTraP2, a program for interpolating band structures and calculating semi-classical transport coefficients. *Comput. Phys. Commun.*, 231, 140.

Mannodi-Kanakkithodi, A. and Chan, M.K. (2022). Accelerated screening of functional atomic impurities in halide perovskites using high-throughput computations and machine learning. *J. Mater. Sci.*, 57, 10736.

Mannodi-Kanakkithodi, A., Toriyama, M.Y., Sen, F.G., Davis, M.J., Klie, R.F., Chan, M.K. (2020). Machine-learned impurity level prediction for semiconductors: The example of Cd-based chalcogenides. *npj Comput. Mater.*, 6(1), 1.

Mannodi-Kanakkithodi, A., Xiang, X., Jacoby, L., Biegaj, R., Dunham, S.T., Gamelin, D.R., Chan, M.K. (2022). Universal machine learning framework for defect predictions in zinc blende semiconductors. *Patterns*, 3(3), 100450.

Margine, E.R. and Giustino, F. (2013). Anisotropic Migdal-Eliashberg theory using Wannier functions. *Phys. Rev. B*, 87(2), 024505.

Marzari, N. and Vanderbilt, D. (1997). Maximally localized generalized Wannier functions for composite energy bands. *Phys. Rev. B*, 56(20), 12847.

Marzari, N., Mostofi, A.A., Yates, J.R., Souza, I., Vanderbilt, D. (2012). Maximally localized Wannier functions: Theory and applications. *Rev. Mod. Phys.*, 84(4), 1419.

May, A.F. and Snyder, G.J. (2017). Introduction to modeling thermoelectric transport at high temperatures. In *Materials, Preparation, and Characterization in Thermoelectrics*, Rowe, D.M. (ed.). CRC Press, Boca Raton.

Mayeshiba, T., Wu, H., Angsten, T., Kaczmarowski, A., Song, Z., Jenness, G., Xie, W., Morgan, D. (2017). The MAterials Simulation Toolkit (MAST) for atomistic modeling of defects and diffusion. *Comp. Mater. Sci.*, 126, 90.

Miller, S.A., Gorai, P., Aydemir, U., Mason, T.O., Stevanović, V., Toberer, E.S., Snyder, G.J. (2017a). SnO as a potential oxide thermoelectric candidate. *J. Mater. Chem. C*, 5(34), 8854.

Miller, S.A., Gorai, P., Ortiz, B.R., Goyal, A., Gao, D., Barnett, S.A., Mason, T.O., Snyder, G.J., Lv, Q., Stevanović, V. et al. (2017b). Capturing anharmonicity in a lattice thermal conductivity model for high-throughput predictions. *Chem. Mater.*, 29(6), 2494.

Miller, S.A., Dylla, M., Anand, S., Gordiz, K., Snyder, G.J., Toberer, E.S. (2018). Empirical modeling of dopability in diamond-like semiconductors. *npj Comput. Mater.*, 4(1), 1.

Miyata, M., Ozaki, T., Takeuchi, T., Nishino, S., Inukai, M., Koyano, M. (2018). High-throughput screening of sulfide thermoelectric materials using electron transport calculations with OpenMX and BoltzTraP. *J. Electron. Mater.*, 47(6), 3254.

Miyazaki, H., Tamura, T., Mikami, M., Watanabe, K., Ide, N., Ozkendir, O.M., Nishino, Y. (2021). Machine learning based prediction of lattice thermal conductivity for half-Heusler compounds using atomic information. *Sci. Rep.*, 11(1), 1.

Na, G.S., Jang, S., Chang, H. (2021). Predicting thermoelectric properties from chemical formula with explicitly identifying dopant effects. *npj Comput. Mater.*, 7(1), 1.

Naik, M.H. and Jain, M. (2018). CoFFEE: Corrections for formation energy and eigenvalues for charged defect simulations. *Comput. Phys. Commun.*, 226, 114.

Naithani, H. and Dasgupta, T. (2019). Critical analysis of single band modeling of thermoelectric materials. *ACS Appl. Energy Mater.*, 3(3), 2200.

Nielsen, M.D., Ozolins, V., Heremans, J.P. (2013). Lone pair electrons minimize lattice thermal conductivity. *Energ. Environ. Sci.*, 6(2), 570.

Noffsinger, J., Giustino, F., Malone, B.D., Park, C.-H., Louie, S.G., Cohen, M.L. (2010). EPW: A program for calculating the electron–phonon coupling using maximally localized Wannier functions. *Comput. Phys. Commun.*, 181(12), 2140.

Ohno, S., Imasato, K., Anand, S., Tamaki, H., Kang, S.D., Gorai, P., Sato, H.K., Toberer, E.S., Kanno, T., Snyder, G.J. (2018). Phase boundary mapping to obtain n-type Mg_3Sb_2-based thermoelectrics. *Joule*, 2(1), 141.

Oliynyk, A.O., Sparks, T.D., Gaultois, M.W., Ghadbeigi, L., Mar, A. (2016). $Gd_{12}Co_{5.3}Bi$ and $Gd_{12}Co_5Bi$, crystalline doppelgänger with low thermal conductivities. *Inorg. Chem.*, 55(13), 6625.

Opahle, I., Parma, A., McEniry, E.J., Drautz, R., Madsen, G.K.H. (2013). High-throughput study of the structural stability and thermoelectric properties of transition metal silicides. *New J. Phys.*, 15(10), 105010.

Ortiz, B.R., Gordiz, K., Gomes, L.C., Braden, T., Adamczyk, J.M., Qu, J., Ertekin, E., Toberer, E.S. (2019). Carrier density control in $Cu_2HgGeTe_4$ and discovery of Hg_2GeTe_4 via phase boundary mapping. *J. Mater. Chem. A*, 7(2), 621.

Ouyang, R., Curtarolo, S., Ahmetcik, E., Scheffler, M., Ghiringhelli, L.M. (2018). SISSO: A compressed-sensing method for identifying the best low-dimensional descriptor in an immensity of offered candidates. *Phys. Rev. Mater.*, 2(8), 083802.

Pan, S., Wang, C., Zhang, Q., Yang, B., Cao, Y., Liu, L., Jiang, Y., You, L., Guo, K., Zhang, J. et al. (2019). $A_2Cu_3In_3Te_8$ (A = Cd, Zn, Mn, Mg): A type of thermoelectric material with complex diamond-like structure and low lattice thermal conductivities. *ACS Appl. Energy Mater.*, 2(12), 8956.

Pan, S., Liu, L., Li, Z., Yan, X., Wang, C., Guo, K., Yang, J., Jiang, Y., Luo, J., Zhang, W. (2021). Embedded in-situ nanodomains from chemical composition fluctuation in thermoelectric $A_2Cu_3In_3Te_8$ (A = Zn, Cd). *Mater. Today Phys.*, 17, 100333.

Park, J., Xia, Y., Ozoliņš, V. (2019). High thermoelectric power factor and efficiency from a highly dispersive band in Ba_2BiAu. *Phys. Rev. Appl.*, 11(1), 014058.

Park, J., Xia, Y., Ganose, A.M., Jain, A., Ozoliņš, V. (2020). High thermoelectric performance and defect energetics of multipocketed full Heusler compounds. *Phys. Rev. Appl.*, 14(2), 024064.

Parker, D. and Singh, D.J. (2010). High-temperature thermoelectric performance of heavily doped PbSe. *Phys. Rev. B*, 82(3), 035204.

Parker, D. and Singh, D.J. (2014). High temperature thermoelectric properties of rock-salt structure PbS. *Solid State Commun.*, 182, 34.

Parker, D., Singh, D.J., Zhang, Q., Ren, Z. (2012). Thermoelectric properties of n-type PbSe revisited. *J. Appl. Phys.*, 111(12), 123701.

Parker, D., Chen, X., Singh, D.J. (2013a). High three-dimensional thermoelectric performance from low-dimensional bands. *Phys. Rev. Lett.*, 110(14), 146601.

Parker, D., May, A.F., Wang, H., McGuire, M.A., Sales, B.C., Singh, D.J. (2013b). Electronic and thermoelectric properties of CoSbS and FeSbS. *Phys. Rev. B*, 87(4), 045205.

Péan, E., Vidal, J., Jobic, S., Latouche, C. (2017). Presentation of the PyDEF post-treatment Python software to compute publishable charts for defect energy formation. *Chem. Phys. Lett.*, 671, 124.

Pei, Y., LaLonde, A.D., Wang, H., Snyder, G.J. (2012). Low effective mass leading to high thermoelectric performance. *Energ. Environ. Sci.*, 5(7), 7963.

Pei, Y., Gibbs, Z.M., Gloskovskii, A., Balke, B., Zeier, W.G., Snyder, G.J. (2014). Optimum carrier concentration in n-type PbTe thermoelectrics. *Adv. Energy Mater.*, 4(13), 1400486.

Peng, H., Scanlon, D.O., Stevanovic, V., Vidal, J., Watson, G.W., Lany, S. (2013). Convergence of density and hybrid functional defect calculations for compound semiconductors. *Phys. Rev. B*, 88, 115201.

Pizzi, G., Volja, D., Kozinsky, B., Fornari, M., Marzari, N. (2014). BoltzWann: A code for the evaluation of thermoelectric and electronic transport properties with a maximally-localized Wannier functions basis. *Comput. Phys. Commun.*, 185(1), 422.

Pöhls, J.-H., Faghaninia, A., Petretto, G., Aydemir, U., Ricci, F., Li, G., Wood, M., Ohno, S., Hautier, G., Snyder, G.J. et al. (2017). Metal phosphides as potential thermoelectric materials. *J. Mater. Chem. C*, 5(47), 12441.

Pöhls, J.-H., Luo, Z., Aydemir, U., Sun, J.-P., Hao, S., He, J., Hill, I.G., Hautier, G., Jain, A., Zeng, X. et al. (2018). First-principles calculations and experimental studies of XYZ_2 thermoelectric compounds: Detailed analysis of van der Waals interactions. *J. Mater. Chem. A*, 6(40), 19502.

Pöhls, J.-H., Chanakian, S., Park, J., Ganose, A.M., Dunn, A., Friesen, N., Bhattacharya, A., Hogan, B., Bux, S., Jain, A. et al. (2021). Experimental validation of high thermoelectric performance in $RECuZnP_2$ predicted by high-throughput DFT calculations. *Mater. Horiz.*, 8(1), 209.

Polak, M.P., Jacobs, R., Mannodi-Kanakkithodi, A., Chan, M.K., Morgan, D. (2022). Machine learning for impurity charge-state transition levels in semiconductors from elemental properties using multi-fidelity datasets. *J. Chem. Phys.*, 156(11), 114110.

Poncé, S., Margine, E.R., Verdi, C., Giustino, F. (2016). EPW: Electron–phonon coupling, transport and superconducting properties using maximally localized Wannier functions. *Comput. Phys. Commun.*, 209, 116.

Qu, J., Stevanović, V., Ertekin, E., Gorai, P. (2020). Doping by design: Finding new n-type dopable ABX_4 Zintl phases for thermoelectrics. *J. Mater. Chem. A*, 8(47), 25306.

Qu, J., Porter, C.E., Gomes, L.C., Adamczyk, J.M., Toriyama, M.Y., Ortiz, B.R., Toberer, E.S., Ertekin, E. (2021). Controlling thermoelectric transport via native defects in the diamond-like semiconductors $Cu_2HgGeTe_4$ and Hg_2GeTe_4. *J. Mater. Chem. A*, 9(46), 26189.

Rahim, W., Skelton, J.M., Scanlon, D.O. (2021). Ca_4Sb_2O and Ca_4Bi_2O: Two promising mixed-anion thermoelectrics. *J. Mater. Chem. A*, 9(36), 20417.

Ravich, I.I. (2013). *Semiconducting Lead Chalcogenides*, vol. 5. Springer Science & Business Media, New York.

Ricci, F., Chen, W., Aydemir, U., Snyder, G.J., Rignanese, G.-M., Jain, A., Hautier, G. (2017). An ab initio electronic transport database for inorganic materials. *Sci. Data*, 4(1), 1.

Scheidemantel, T., Ambrosch-Draxl, C., Thonhauser, T., Badding, J., Sofo, J. (2003). Transport coefficients from first-principles calculations. *Phys. Rev. B*, 68(12), 125210.

Schulz, W.W., Allen, P.B., Trivedi, N. (1992). Hall coefficient of cubic metals. *Phys. Rev. B*, 45(19), 10886.

Schweika, W., Hermann, R., Prager, M., Persson, J., Keppens, V. (2007). Dumbbell rattling in thermoelectric zinc antimony. *Phys. Rev. Lett.*, 99(12), 125501.

Seko, A., Togo, A., Hayashi, H., Tsuda, K., Chaput, L., Tanaka, I. (2015). Prediction of low-thermal-conductivity compounds with first-principles anharmonic lattice-dynamics calculations and Bayesian optimization. *Phys. Rev. Lett.*, 115(20), 205901.

Sevik, C. and Çağın, T. (2009). Assessment of thermoelectric performance of Cu_2ZnSnX_4, X = S, Se, and Te. *Appl. Phys. Lett.*, 95(11), 112105.

Sheng, Y., Wu, Y., Yang, J., Lu, W., Villars, P., Zhang, W. (2020). Active learning for the power factor prediction in diamond-like thermoelectric materials. *npj Comput. Mater.*, 6(1), 1.

Singh, D.J. (2010). Doping-dependent thermopower of PbTe from Boltzmann transport calculations. *Phys. Rev. B*, 81(19), 195217.

Sjakste, J., Vast, N., Calandra, M., Mauri, F. (2015). Wannier interpolation of the electron–phonon matrix elements in polar semiconductors: Polar-optical coupling in GaAs. *Phys. Rev. B*, 92(5), 054307.

Skoug, E.J. and Morelli, D.T. (2011). Role of lone-pair electrons in producing minimum thermal conductivity in nitrogen-group chalcogenide compounds. *Phys. Rev. Lett.*, 107(23), 235901.

Skoug, E.J., Cain, J.D., Morelli, D.T. (2010). Structural effects on the lattice thermal conductivity of ternary antimony- and bismuth-containing chalcogenide semiconductors. *Appl. Phys. Lett.*, 96(18), 181905.

Snyder, G.J. and Toberer, E.S. (2008). Complex thermoelectric materials. *Nature Materials*, 7(2), 105.

Sofo, J.O. and Mahan, G. (1994). Optimum band gap of a thermoelectric material. *Phys. Rev. B*, 49(7), 4565.

Song, Q., Liu, T.-H., Zhou, J., Ding, Z., Chen, G. (2017). Ab initio study of electron mean free paths and thermoelectric properties of lead telluride. *Mater. Today Phys.*, 2, 69.

Souza, I., Marzari, N., Vanderbilt, D. (2001). Maximally localized Wannier functions for entangled energy bands. *Phys. Rev. B*, 65(3), 035109.

Stoliaroff, A., Jobic, S., Latouche, C. (2018). PyDEF 2.0: An easy to use post-treatment software for publishable charts featuring a graphical user interface. *J. Comput. Chem.*, 39(26), 2251.

Sun, J. and Singh, D.J. (2016). Thermoelectric properties of n-type $SrTiO_3$. *APL Mater.*, 4(10), 104803.

Tadano, T., Gohda, Y., Tsuneyuki, S. (2014). Anharmonic force constants extracted from first-principles molecular dynamics: Applications to heat transfer simulations. *J. Phys. Condens. Mat.*, 26(22), 225402.

Takagiwa, Y., Hou, Z., Tsuda, K., Ikeda, T., Kojima, H. (2021). Fe–Al–Si thermoelectric (FAST) materials and modules: Diffusion couple and machine-learning-assisted materials development. *ACS Appl. Mater. Interfaces*, 13(45), 53346.

Tan, G., Hao, S., Zhao, J., Wolverton, C., Kanatzidis, M.G. (2017). High thermoelectric performance in electron-doped $AgBi_3S_5$ with ultralow thermal conductivity. *J. Am. Chem. Soc.*, 139(18), 6467.

Terada, Y., Ohkubo, K., Mohri, T., Suzuki, T. (2002). Thermal conductivity of intermetallic compounds with metallic bonding. *Mater. Trans.*, 43(12), 3167.

Tewari, A., Dixit, S., Sahni, N., Bordas, S.P. (2020). Machine learning approaches to identify and design low thermal conductivity oxides for thermoelectric applications. *Data Centric Eng.*, 1, e8.

Toberer, E.S., May, A.F., Scanlon, C.J., Snyder, G.J. (2009). Thermoelectric properties of p-type LiZnSb: Assessment of ab initio calculations. *J. Appl. Phys.*, 105(6), 063701.

Togo, A., Chaput, L., Tanaka, I. (2015). Distributions of phonon lifetimes in Brillouin zones. *Phys. Rev. B*, 91(9), 094306.

Toriyama, M.Y., Qu, J., Snyder, G.J., Gorai, P. (2021). Defect chemistry and doping of BiCuSeO. *J. Mater. Chem. A*, 9(36), 20685.

Toriyama, M.Y., Brod, M.K., Gomes, L.C., Bipasha, F.A., Assaf, B.A., Ertekin, E., Snyder, G.J. (2022a). Tuning valley degeneracy with band inversion. *J. Mater. Chem. A*, 10(3), 1588.

Toriyama, M.Y., Carranco, A.N., Snyder, G.J., Gorai, P. (2022b). Material descriptors to predict thermoelectric performance of narrow-gap semiconductors and semimetals. *ChemRxiv*. doi: 10.26434/chemrxiv-2022-rm5ll.

Toriyama, M.Y., Cheikh, D., Bux, S.K., Snyder, G.J., Gorai, P. (2022c). Y_2Te_3: A new n-type thermoelectric material. *ACS Appl. Mater. Interfaces*, 14(38), 43517.

Tranås, R., Løvvik, O.M., Tomic, O., Berland, K. (2022). Lattice thermal conductivity of half-Heuslers with density functional theory and machine learning: Enhancing predictivity by active sampling with principal component analysis. *Comp. Mater. Sci.*, 202, 110938.

Tsuda, N., Nasu, K., Fujimori, A., Siratori, K. (2013). *Electronic Conduction in Oxides*. vol. 94. Springer Science & Business Media, New York.

Vasiliev, A., Ivanov, O., Zhezhu, M., Yapryntsev, M. (2021). Synthesis and properties of thermoelectric nanomaterial $AgInSe_2$ with a chalcopyrite structure. *Nanobiotech. Rep.*, 16(3), 357.

Verdi, C. and Giustino, F. (2015). Fröhlich electron–phonon vertex from first principles. *Phys. Rev. Lett.*, 115(17), 176401.

Walukiewicz, W. (2001). Intrinsic limitations to the doping of wide-gap semiconductors. *Physica B*, 302, 123.

Wang, Y., Chen, X., Cui, T., Niu, Y., Wang, Y., Wang, M., Ma, Y., Zou, G. (2007). Enhanced thermoelectric performance of PbTe within the orthorhombic *pnma* phase. *Phys. Rev. B*, 76(15), 155127.

Wang, H., Pei, Y., LaLonde, A.D., Snyder, G.J. (2011). Heavily doped *p*-type PbSe with high thermoelectric performance: An alternative for PbTe. *Adv. Mater.*, 23(11), 1366.

Wang, H., Gibbs, Z.M., Takagiwa, Y., Snyder, G.J. (2014). Tuning bands of PbSe for better thermoelectric efficiency. *Energ. Environ. Sci.*, 7(2), 804.

Wang, H., Qin, G., Qin, Z., Li, G., Wang, Q., Hu, M. (2018a). Lone-pair electrons do not necessarily lead to low lattice thermal conductivity: An exception of two-dimensional penta-CN_2. *J. Phys. Chem. Lett.*, 9(10), 2474.

Wang, X., Witkoske, E., Maassen, J., Lundstrom, M. (2018b). LanTraP: A code for calculating thermoelectric transport properties with the Landauer formalism. *arXiv preprint*. arXiv:1806.08888.

Wang, T., Xiong, Y., Wang, Y., Qiu, P., Song, Q., Zhao, K., Yang, J., Xiao, J., Shi, X., Chen, L. (2020a). Cu_3ErTe_3: A new promising thermoelectric material predicated by high-throughput screening. *Mater. Today Phys.*, 12, 100180.

Wang, X., Zeng, S., Wang, Z., Ni, J. (2020b). Identification of crystalline materials with ultra-low thermal conductivity based on machine learning study. *J. Phys. Chem. C*, 124(16), 8488.

West, D., Sun, Y., Wang, H., Bang, J., Zhang, S. (2012). Native defects in second-generation topological insulators: Effect of spin-orbit interaction on Bi_2Se_3. *Phys. Rev. B*, 86(12), 121201.

Wood, M., Toriyama, M.Y., Dugar, S., Male, J., Anand, S., Stevanović, V., Snyder, G.J. (2021). Phase boundary mapping of tin-doped ZnSb reveals thermodynamic route to high thermoelectric efficiency. *Adv. Energy Mater.*, 11(20), 2100181.

Xi, L., Pan, S., Li, X., Xu, Y., Ni, J., Sun, X., Yang, J., Luo, J., Xi, J., Zhu, W. et al. (2018). Discovery of high-performance thermoelectric chalcogenides through reliable high-throughput material screening. *J. Am. Chem. Soc.*, 140(34), 10785.

Xia, Y. (2018). Revisiting lattice thermal transport in PbTe: The crucial role of quartic anharmonicity. *Appl. Phys. Lett.*, 113(7), 073901.

Xia, Y., Pal, K., He, J., Ozoliņš, V., Wolverton, C. (2020). Particlelike phonon propagation dominates ultralow lattice thermal conductivity in crystalline Tl_3VSe_4. *Phys. Rev. Lett.*, 124(6), 065901.

Xing, G., Sun, J., Li, Y., Fan, X., Zheng, W., Singh, D.J. (2017). Electronic fitness function for screening semiconductors as thermoelectric materials. *Phys. Rev. Mater.*, 1(6), 065405.

Xu, L., Zheng, Y., Zheng, J.-C. (2010). Thermoelectric transport properties of PbTe under pressure. *Phys. Rev. B*, 82(19), 195102.

Xu, L., Wang, H.-Q., Zheng, J.-C. (2011). Thermoelectric properties of PbTe, SnTe, and GeTe at high pressure: An ab initio study. *J. Electron. Mater.*, 40(5), 641.

Xu, B., Di Gennaro, M., Verstraete, M.J. (2020). Thermoelectric properties of elemental metals from first-principles electron–phonon coupling. *Phys. Rev. B*, 102(15), 155128.

Xue, Q., Liu, H., Fan, D., Cheng, L., Zhao, B., Shi, J. (2016). LaPtSb: A half-Heusler compound with high thermoelectric performance. *Phys. Chem. Chem. Phys.*, 18(27), 17912.

Yahyaoglu, M., Soldi, T., Ozen, M., Candolfi, C., Snyder, G.J., Aydemir, U. (2021). Stress/pressure-stabilized cubic polymorph of Li_3Sb with improved thermoelectric performance. *J. Mater. Chem. A*, 9(44), 25024.

Yan, J., Gorai, P., Ortiz, B., Miller, S., Barnett, S.A., Mason, T., Stevanović, V., Toberer, E.S. (2015). Material descriptors for predicting thermoelectric performance. *Energ. Environ. Sci.*, 8(3), 983.

Yang, J., Li, H., Wu, T., Zhang, W., Chen, L., Yang, J. (2008). Evaluation of half-Heusler compounds as thermoelectric materials based on the calculated electrical transport properties. *Adv. Funct. Mater.*, 18(19), 2880.

Yang, X., Dai, Z., Zhao, Y., Liu, J., Meng, S. (2018). Low lattice thermal conductivity and excellent thermoelectric behavior in Li_3Sb and Li_3Bi. *J. Phys. Condens. Mat.*, 30(42), 425401.

Ye, X., Feng, Z., Xu, Y., Jian, M., Yan, Y., Zhang, Y., Zhao, G. (2021). A theoretical study on the thermal conductivity and thermoelectric properties of CoNbSi and CoNbSn. *J. Phys. Chem. C*, 125(18), 10068.

You, Y., Su, X., Hao, S., Liu, W., Yan, Y., Zhang, T., Zhang, M., Wolverton, C., Kanatzidis, M.G., Tang, X. (2018). Ni and Se co-doping increases the power factor and thermoelectric performance of CoSbS. *J. Mater. Chem. A*, 6(31), 15123.

Yu, P. and Cardona, M. (2010). *Fundamentals of Semiconductors*. Springer Science & Business Media, Berlin/Heidelberg.

Zhang, S., Wei, S.-H., Zunger, A. (1999). Overcoming doping bottlenecks in semiconductors and wide-gap materials. *Physica B*, 273–274, 976.

Zhang, Q., Wang, H., Liu, W., Wang, H., Yu, B., Zhang, Q., Tian, Z., Ni, G., Lee, S., Esfarjani, K. et al. (2012a). Enhancement of thermoelectric figure-of-merit by resonant states of aluminium doping in lead selenide. *Energ. Environ. Sci.*, 5(1), 5246.

Zhang, Y., Skoug, E., Cain, J., Ozoliņš, V., Morelli, D., Wolverton, C. (2012b). First-principles description of anomalously low lattice thermal conductivity in thermoelectric Cu–Sb–Se ternary semiconductors. *Phys. Rev. B*, 85(5), 054306.

Zhang, R.-Z., Chen, K., Du, B., Reece, M.J. (2017). Screening for Cu–S based thermoelectric materials using crystal structure features. *J. Mater. Chem. A*, 5(10), 5013.

Zhang, Q., Xi, L., Zhang, J., Wang, C., You, L., Pan, S., Guo, K., Li, Z., Luo, J. (2021). Influence of Ag substitution on thermoelectric properties of the quaternary diamond-like compound $Zn_2Cu_3In_3Te_8$. *J. Materiomics*, 7(2), 236.

Zhao, L.-D., He, J., Berardan, D., Lin, Y., Li, J.-F., Nan, C.-W., Dragoe, N. (2014). BiCuSeO oxyselenides: New promising thermoelectric materials. *Energ. Environ. Sci.*, 7(9), 2900.

Zhong, Y., Sarker, D., Fan, T., Xu, L., Li, X., Qin, G.-Z., Han, Z.-K., Cui, J. (2021). Computationally guided synthesis of high performance thermoelectric materials: Defect engineering in $AgGaTe_2$. *Adv. Electron. Mater.*, 7(4), 2001262.

Zhou, J.-J. and Bernardi, M. (2016). Ab initio electron mobility and polar phonon scattering in GaAs. *Phys. Rev. B*, 94(20), 201201.

Zhou, X. and Zhang, Z. (2020). Electron–phonon coupling in $CsPbBr_3$. *AIP Adv.*, 10(12), 125015.

Zhou, F., Nielson, W., Xia, Y., Ozoliņš, V. (2014). Lattice anharmonicity and thermal conductivity from compressive sensing of first-principles calculations. *Phys. Rev. Lett.*, 113(18), 185501.

Zhou, J., Zhu, H., Liu, T.-H., Song, Q., He, R., Mao, J., Liu, Z., Ren, W., Liao, B., Singh, D.J. et al. (2018). Large thermoelectric power factor from crystal symmetry-protected non-bonding orbital in half-Heuslers. *Nat. Commun.*, 9(1), 1.

Zhou, F., Nielson, W., Xia, Y., Ozoliņš, V. (2019a). Compressive sensing lattice dynamics I. General formalism. *Phys. Rev. B*, 100(18), 184308.

Zhou, F., Sadigh, B., Åberg, D., Xia, Y., Ozoliņš, V. (2019b). Compressive sensing lattice dynamics II. Efficient phonon calculations and long-range interactions. *Phys. Rev. B*, 100(18), 184309.

Zhou, J.-J., Park, J., Lu, I.-T., Maliyov, I., Tong, X., Bernardi, M. (2021). Perturbo: A software package for ab initio electron–phonon interactions, charge transport and ultrafast dynamics. *Comput. Phys. Commun.*, 264, 107970.

Zhu, H., Hautier, G., Aydemir, U., Gibbs, Z.M., Li, G., Bajaj, S., Pöhls, J.-H., Broberg, D., Chen, W., Jain, A. et al. (2015). Computational and experimental investigation of TmAgTe$_2$ and XYZ$_2$ compounds, a new group of thermoelectric materials identified by first-principles high-throughput screening. *J. Mater. Chem. C*, 3(40), 10554.

Zhu, H., Mao, J., Li, Y., Sun, J., Wang, Y., Zhu, Q., Li, G., Song, Q., Zhou, J., Fu, Y. et al. (2019). Discovery of TaFeSb-based half-Heuslers with high thermoelectric performance. *Nat. Commun.*, 10(1), 1.

Zhu, J., Zhang, X., Guo, M., Li, J., Hu, J., Cai, S., Cai, W., Zhang, Y., Sui, J. (2021a). Restructured single parabolic band model for quick analysis in thermoelectricity. *npj Comput. Mater.*, 7(1), 1.

Zhu, T., He, R., Gong, S., Xie, T., Gorai, P., Nielsch, K., Grossman, J.C. (2021b). Charting lattice thermal conductivity for inorganic crystals and discovering rare earth chalcogenides for thermoelectrics. *Energ. Environ. Sci.*, 14(6), 3559.

PART 2
Thermoelectric Materials

3

Thermoelectric Copper and Silver Chalcogenides

Holger KLEINKE
Department of Chemistry and Waterloo Institute for Nanotechnology, University of Waterloo, Canada

3.1. Introduction

The importance of thermoelectric (TE) energy generation as a sustainable energy conversion method continues to increase (He and Tritt 2017; Shi et al. 2020). Historically, TE generators have been used in a number of spacecraft for several decades (Furlong and Wahlquist 1999; Yang and Caillat 2006). Since the 1990s, TEs have been studied for use in automobiles and some stationary applications to convert their waste heat into useful energy (Yang and Caillat 2006; Yang and Stabler 2009; Matsumoto et al. 2015; Orr et al. 2016). More recently, applications of interest began to include TE generators powered by body heat (Yang et al. 2007; Leonov and Vullers 2009; Francioso et al. 2011; Tian et al. 2019), as well as powering the billions of sensors in the Internet of Things of Society 5.0 (Park et al. 2019; Zaia et al. 2019; Freer and Powell 2020).

The efficiency of the TE energy conversion increases with an increasing figure of merit $zT = T\alpha^2 \sigma \kappa^{-1}$. Here, T denotes the average temperature, α

For a color version of all the figures in this chapter, see www.iste.co.uk/akinaga/ thermoelectric1.zip.

Thermoelectric Micro/Nano Generators 1,
coordinated by Hiroyuki AKINAGA, Atsuko KOSUGA,
Takao MORI and Gustavo ARDILA.
© ISTE Ltd 2023.

denotes the Seebeck coefficient, σ denotes the electrical conductivity and κ denotes the thermal conductivity. The thermal conductivity consists of two parts, the lattice (κ_L) and the electronic thermal conductivity (κ_e): $\kappa = \kappa_L + \kappa_e$. κ_e is proportional to σ and T, as expressed in the Wiedemann–Franz law: $\kappa_e = L_0 \sigma T$, where L_0 is the Lorenz number. With the exception of κ_L, all of these physical properties depend on the charge carrier concentration n. As n increases, the Seebeck coefficient decreases, and σ and κ_e increase. Therefore, κ_L is the only property that can be optimized (decreased) independently, mainly by incorporating heavy elements into TEs (Snyder and Toberer 2008; Kleinke 2010; Shi et al. 2019). This readily explains the history of focusing on variants of Bi_2Te_3 and PbTe (Pei et al. 2011), with their average molar masses of 160.2 g mol^{-1} and 167.4 g mol^{-1} and κ_L values of 1.5 W m^{-1}K^{-1} and 2.2 W m^{-1}K^{-1}, respectively, and later on more complex antimonides and tellurides. Examples for the latter are, in no particular order, the antimonides $Ba_{0.08}La_{0.05}Yb_{0.04}Co_4Sb_{12}$ (Shi et al. 2011), $Yb_{14}Mn_{0.4}Al_{0.6}Sb_{11}$ (Toberer et al. 2008) and β-Zn_4Sb_3 (Caillat et al. 1997), and the tellurides $Tl_9Bi_{0.98}Te_6$ (Guo et al. 2013), $Tl_{8.05}Sn_{1.95}Te_6$ and $Tl_{8.10}Pb_{1.90}Te_6$ (Guo et al. 2014) and $Tl_2Ag_{12}Te_{7.4}$ (Shi et al. 2018), all of which exhibit peak zT values above unity and, in part, κ_L values below 1 W m^{-1}K^{-1}.

Further improvements can be achieved by different nanostructuring methods, which in part resulted in peak zT values above 2. Prominent examples are superlattices of Bi_2Te_3/Sb_2Te_3 (Venkatasubramanian et al. 2001) and $PbSe_{0.98}Te_{0.02}/PbTe$ (Harman et al. 2005), Bi_2Te_3 nanowires (Lv et al. 2013), Na-doped PbTe/SrTe (Biswas et al. 2012) and $Cu_2Se/CuInSe_2$ nanocomposites (Olvera et al. 2017), and nanostructured Cu_2Se (Gahtori et al. 2015).

Recently, the focus has begun to shift from tellurides to more abundant and environmentally benign sulfides and selenides. For these materials, other methods of achieving low thermal conductivity must be used because they are composed of lighter elements. An advantageous property in this regard is anharmonicity, as demonstrated for SnSe (Lee et al. 2019; Zhou et al. 2021) and the tetrahedrites $Cu_{12-x}M_xSb_4S_{13}$, with M = Mn, Fe, Co, Ni, Zn and Hg. A second is extreme disorder, all the way up to a "liquid-like" sublattice as realized in various copper chalcogenides, the focus of this chapter, including β-Cu_2Se (Liu et al. 2012).

3.2. Binary copper and silver chalcogenides

The TE properties of the sulfide Cu_2S were investigated as early as the 19th century (Dennler et al. 2014). The *p*-type selenide Cu_2Se was used in a device in the 1980s, which unfortunately failed rapidly as Cu ions diffused out of the TE leg (Brown et al. 2013). This spectacular failure has hindered the study of these and related copper chalcogenides for decades. But in the last 10 years, more and more research groups have sought to identify methods of inhibiting Cu ion migration to minimize decay (Mayasree et al. 2012; Qiu et al. 2016; Wei et al. 2019), leading to a renewed interest in these materials.

The six Cu and Ag chalcogenides Cu_2Q and Ag_2Q, with Q = S, Se and Te, all adopt structures with fixed positions of the Q^{2-} anions, and Cu^+ and Ag^+ ions in different holes. These binaries undergo various phase transitions, which in part result in superionic conductivity. The chalcocite Cu_2S adopts three different structures at different temperatures. Below 370 K, the monoclinic γ-phase is thermodynamically preferred. At approximately 370 K, a phase transformation occurs in the hexagonal β-phase. At approximately 700 K, the β-phase transforms into the superionic cubic α-phase (Balapanov et al. 2004). Despite its relatively low average molar mass of 53.1 g mol^{-1}, its lattice thermal conductivity κ_L is below 0.8 W m^{-1}K^{-1}, which is a consequence of its liquid-like Cu atom sublattice. The slightly Cu-deficient $Cu_{1.97}S$ shows an exceptionally high peak figure of merit with its zT_{max} = 1.7 at 1,000 K (He et al. 2014), and $Cu_{1.97}S$ produced via a melt-solidification technique has an even higher zT_{max} of 1.9 at 973 K (Zhao et al. 2015a). However, its superionic property leads to a low stability of this material. The addition of excessive amounts of iodine resulted in superionic Cu_2S exhibiting zT_{max} = 1.8 also at 973 K, with the benefit of increased stability (Zhao et al. 2020).

The Cu ion conductor β-Cu_2Se is the thermodynamically preferred modification above 410 K. Its structure is based on fluorite, where the Se atoms form a cubic closed-packed arrangement with Cu atoms in all tetrahedral holes and are kinetically disordered over various interstitial sites (see Figure 3.1).

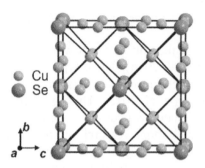

Figure 3.1. *Crystal structure of β-Cu$_2$Se*

The total thermal conductivity values, including the electronic contribution, are mostly below 1 W m^{-1}K^{-1} for Cu$_2$Se-based materials. Several research groups have reported excellent zT_{max} values above 950 K, such as 1.8 after self-propagating synthesis (Su et al. 2014), 1.8 after melt quenching (Zhao et al. 2015b) and finally zT_{max} values between 2 and 2.1 by including defects on the nanoscale (Gahtori et al. 2015), spark plasma sintering of nanopowders (Tafti et al. 2016) and doping Cu$_2$Se with lithium (Cu$_{1.97}$Li$_{0.03}$Se) (Hu et al. 2018) and sodium (Cu$_{1.96}$Na$_{0.04}$Se) (Zhu et al. 2019) with nano- and micropores, respectively. Figure 3.2 shows the figure of merit data of the last three examples along with the respective low thermal conductivity data, including when alloyed with S (Cu$_{1.96}$S$_{0.2}$Se$_{0.8}$) (Mao et al. 2020). The peaks around 400 K are caused by phase transitions.

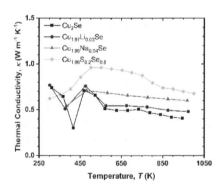
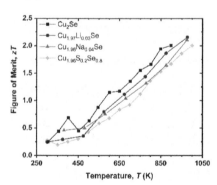

Figure 3.2. *Thermal conductivity (left) and figure of merit (right) of leading Cu$_2$Se materials*

Further achievements were made via the nanocomposite approach, which generally resulted in lower thermal conductivity and higher zT values. The addition of various carbon nanostructures was particularly advantageous. For example, the use of carbon nanodots (CD) (Hu et al. 2020), graphene (Li et al. 2018) and carbon nanotubes (CNT) (Nunna et al. 2017) led to peak zT values between 2.2 and 2.4, all with total thermal conductivity values below 0.6 W m^{-1}K^{-1} (see Figure 3.3).

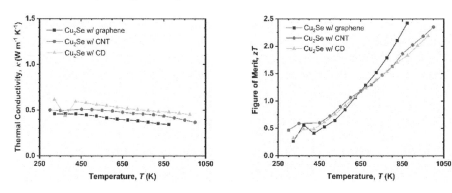

Figure 3.3. *Thermal conductivity (left) and figure of merit (right) of leading Cu$_2$Se nanocomposites*

Contrary to the usual trends, Cu$_2$Te shows less impressive TE properties than its lighter homologues. Of its five modifications, the cubic ε-phase, stable above 850 K, has the largest zT_{max} values, including 1.0 at 900 K when doped with Ag (Ballikaya et al. 2013) and 1.1 at 1,000 K after minimizing the carrier concentration (He et al. 2015).

α-Ag$_2$S transforms into the cubic β-phase at approximately 450 K, which in turn changes into the γ-modification at approximately 860 K (Sharma and Chang 1986). Ag$_2$S falls behind the binary copper sulfides in TE performance, reaching a more modest peak zT value of 0.55 at 580 K (Wang et al. 2019).

On the contrary, the heavier binary silver chalcogenides, Ag$_2$Se and Ag$_2$Te, reach their peak zT values at significantly lower temperatures, in line with their narrower band gaps of, for example, $E_g < 0.2$ eV for Ag$_2$Se (Ferhat and Nagao 2000), compared to Ag$_2$S and the copper analogs with gaps of 1.0–1.2 eV. Ag_2Se shows a phase transition around 410 K from an orthorhombic modification to a cubic ion conductor (Billetter and Ruschewitz 2008), and reaches zT values above 1 at relatively low temperatures, for example 1.2 at 390 K with a low carrier concentration of $n = 10^{18}$ cm^{-3} (Yang et al. 2017). At approximately room temperature, Ag_2Se showed zT values above 0.8, making it an attractive alternative to Bi_2Te_3 (Huang et al. 2021).

Ag_2Te is – thus far – the best performing binary silver chalcogenide. Its monoclinic α-phase is preferred below 418 K, the cubic β-phase above 418 K and the γ-phase above 1,075 K (Fujikane et al. 2005). Its Sb-doped variant was reported to achieve $zT_{max} = 1.4$ at 410 K (Zhu et al. 2020).

3.3. Ternary and higher copper and silver chalcogenides

3.3.1. *Minerals based on copper and silver chalcogenides*

Several minerals in this class, mostly based on copper sulfides, are under investigation for their outstanding TE properties. The high-temperature phase of the bornite Cu_5FeS_4 can be viewed as four units of Cu_2S (i.e. Cu_8S_4), where one Cu is replaced by Fe and two Cu atoms by vacancies □, resulting in $Cu_5Fe□_2S_4$. Its low lattice thermal conductivity of $\kappa_L = 0.5$ W m^{-1}K^{-1} is a consequence of the deficiencies and Cu ion disorder (Qiu et al. 2014). The Fe atoms reduce the migration of Cu ions and increase the stability of bornite. The variant $Cu_{4.972}Fe_{0.968}S_4$ has been reported to reach $zT_{max} = 0.79$ at 550 K (Long et al. 2018).

The colusites $Cu_{26}M_2E_6S_{32}$ (with M = V, Nb, Ta, Cr, Mo, W and E = Ge, Sn, As, Sb) also have low thermal conductivity, which is a consequence of their large complex unit cells. Impressive zT_{max} values have been recently reported for $Cu_{26}V_2Sn_6S_{32}$ with 0.9 at 700 K (Bourgès et al. 2018), for $Cu_{26}Cr_2Ge_6S_{32}$ also with 0.9 at 700 K (Pavan Kumar et al. 2019), as well as for $Cu_{26}Ta_2Sn_{5.5}S_{32}$ and $Cu_{26}Nb_2Ge_6S_{32}$, both with $zT_{max} = 1.0$ at 670 K (Bouyrie et al. 2017).

The tetrahedrites $Cu_{12-x}M_xSb_4S_{13}$ (M = Mn, Fe, Co, Ni, Zn, Hg) have attracted the most attention among the copper sulfide minerals. In addition to their large complex unit cells, one Cu atom is bonded to only three S atoms instead of the more common four S atoms in a tetrahedral conformation, and the Sb–S polyhedra are highly distorted because of the lone pair effect of

Sb^{3+}. The resulting large anharmonicity reflects itself in a low lattice thermal conductivity of typically below 0.6 W m^{-1}K^{-1}. With x = 2, $Cu_{12-x}Zn_xSb_4S_{13}$ is an intrinsic semiconductor and a *p*-doped semiconductor when x < 2. $Cu_{11}MnSb_4S_{13}$ was the first tetrahedrite to surpass a *zT* of unity, with zT_{max} = 1.1 at 575 K (Heo et al. 2014).

3.3.2. Tl-containing copper and silver chalcogenides

The first representative of a thallium silver chalcogenide with high TE performance was $TlAg_9Te_5$ with zT_{max} = 1.2 at 700 K (Kurosaki et al. 2005a). This material crystallizes in a complex unit cell of large hexagonal channels formed by the Ag and Te atoms filled with chains of Tl atoms. The Ag sites are highly disordered with occupancies between 46% and 88% (Paccard et al. 1992), leading to enhanced phonon scattering and thus lower thermal conductivity.

Similar channels exist in the structure of $Tl_2Ag_{12-x}Te_{7.4}$, which is a variant of the $Zr_2Fe_{12}P_7$ type. Therein, Ag site occupancies are generally higher than in $TlAg_9Te_5$. In contrast to $TlAg_9Te_5$, $Tl_2Ag_{12-x}Te_{7.4}$ forms a composite structure, where composite A is the $Tl_2Ag_{12}Te_6$ framework and composite B is an incommensurate Te atom chain with ~1.4 Te atoms per $Tl_2Ag_{12}Te_6$ unit. This linear Te chain consists of a modulation of Te_2^{2-} and Te_3^{4-} units, which in turn also distorts the neighboring Ag atoms. Finally, the composite structure with its distortions and deficiencies leads to an ultralow thermal conductivity of < 0.3 W m^{-1}K^{-1}, and a peak *zT* value of 1.1 at 520 K. In both cases, $TlAg_9Te_5$ and $Tl_2Ag_{12-x}Te_{7.4}$, three-dimensionally extended Ag–Ag contacts < 3.3 Å are present. In addition to the site deficiencies, Ag ion conductivity cannot be excluded, but reproducibility measurements revealed no stability problems in the case of $Tl_2Ag_{12-x}Te_{7.4}$ (Shi et al. 2018).

3.3.3. Ba-containing copper and silver chalcogenides

Twenty years ago, the low-temperature TE properties of $A_2BaCu_8Te_{10}$ (with *A* = K, Rb, Cs) were determined (Patschke et al. 2001). Its complex crystal structure consists of Ba-filled pentagonal dodecahedral Cu_8Te_{12} cages, Te^{2-} anions, as well as Te_2^{2-} dumbbells and *A* cations with enlarged anisotropic displacement parameters. $Rb_2BaCu_8Te_{10}$ has an optical band gap of E_g = 0.28 eV and a low thermal conductivity of κ = 1.4 W m^{-1}K^{-1} at 300 K.

Also in 2001, the first ternary barium copper telluride was introduced, namely $BaCu_2Te_2$, and its low-temperature TE properties were presented (Wang and DiSalvo 2001). $BaCu_2Te_2$ is isostructural with α-$BaCu_2S_2$ and $BaCu_2Se_2$, with double chains of edge-sharing $CuTe_4$ tetrahedra running along the b axis. These columns are interconnected via corner sharing to form a three-dimensional network of $CuTe_4$ tetrahedra with Ba cations in their channels (see Figure 3.4). The as-prepared $BaCu_2Te_2$ is a degenerate semiconductor, indicating unidentified Cu atom deficiencies, with a thermal conductivity of 2.3 W $m^{-1}K^{-1}$ at 300 K.

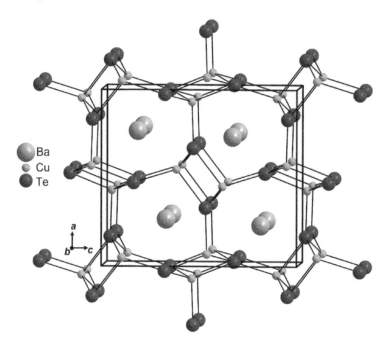

Figure 3.4. *Crystal structure of $BaCu_2Te_2$*

Over a decade later, $BaCu_2Te_2$ was shown to reach a peak zT value of 0.72 at 823 K, and $BaCu_{1.9}Ag_{0.1}Te_2$ a zT_{max} value of 1.1 at 823 K, partly due to its lower thermal conductivity of 0.6 W $m^{-1}K^{-1}$ at 300 K and 0.5 W $m^{-1}K^{-1}$ at 823 K (Yang et al. 2019). The addition of more Cu to this material, corresponding to a nominal formula of $BaCu_{2.04}Te_2$, lowered the room temperature thermal conductivity from 0.9 W $m^{-1}K^{-1}$ to 0.7 W $m^{-1}K^{-1}$ after the formation of nano-precipitates with a size of 20–50 nm. Therefore, a superior $zT_{max} = 1.3$ is achieved at 833 K (Guo et al. 2020).

The corresponding sulfide $BaCu_2S_2$, which in its β-modification adopts the $ThCr_2Si_2$ type at elevated temperatures, is less competitive. After doping with K, it reaches a modest zT_{max} of 0.28 at 820 K (Kurosaki et al. 2005b). More recently, the doping of the selenide $BaCu_2Se_2$ with sodium yielded a competitive zT_{max} of 1.0 at 773 K with a thermal conductivity of approximately 1 W m^{-1}K^{-1} at room temperature (Li et al. 2015).

On the contrary, the as-prepared Ag telluride, $BaAg_2Te_2$, is an intrinsic semiconductor with low electrical conductivity. Conductivity increases in the solid solution of $BaCu_xAg_{2-x}Te_2$ with increasing x (Cu concentration), presumably due to the increasing Cu atom deficiencies (Assoud et al. 2008). Superior performance was achieved at a Cu:Ag ratio of 1:1, corresponding to $BaCuAgTe_2$, with its $zT_{max} = 1.3$ at 823 K (Tang et al. 2022).

A more complex, low symmetry structure is adopted by $Ba_3Cu_{14-x}Te_{12}$, which forms its own structure type (Assoud et al. 2006). As with $BaCu_2Te_2$, the structure consists of a three-dimensionally extended network of Cu–Te polyhedra. In addition to the common $CuTe_4$ tetrahedra of $BaCu_2Te_2$, $Ba_3Cu_{14-x}Te_{12}$ also contains $CuTe_3$ triangles as well as Te_2^{2-} dumbbells (see Figure 3.5).

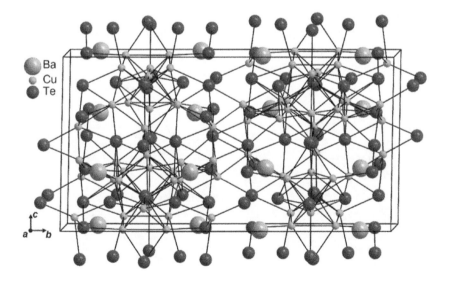

Figure 3.5. *Crystal structure of $Ba_3Cu_{14-x}Te_{12}$*

$Ba_3Cu_{14-x}Te_{12}$ is an electron-precise material for x = 0, as can be seen from the formula $(Ba^{2+})_3(Cu^+)_{14}(Te_2^{2-})_2(Te^{2-})_8$. With always x > 0 and x as large as 0.875, $Ba_3Cu_{14-x}Te_{12}$ exhibits extrinsic p-type semiconducting properties. Like many of the materials discussed here, this material has a low thermal conductivity below 1.0 W m^{-1}K^{-1}, but suffers from stability issues as a result of Cu ion migration at elevated temperatures. Therefore, repeat measurements result in non-reproducible data. In addition, relatively low Seebeck values occur with a small figure of merit, with zT_{max} of the order of only 0.1 (Sturm et al. 2021).

The addition of sulfur or selenium to the mixture has resulted in a number of unprecedented crystal structures, including $Ba_2Cu_{4-x}Se_yTe_{5-y}$ (Mayasree et al. 2010), $Ba_3Cu_{16-x}S_{11-y}Te_y$ (Kuropatwa et al. 2011) and $Ba_3Cu_{16-x}Se_{11-y}Te_y$ (Kuropatwa et al. 2009), and $BaCu_{6-x}STe_6$ and $BaCu_{6-x}Se_{1-y}Te_{6+y}$ (Mayasree et al. 2011). With the exception of $Ba_2Cu_{4-x}Se_yTe_{5-y}$, the TE properties of all of these sulfide-tellurides and selenide-tellurides have recently been determined.

$Ba_3Cu_{16-x}S_{11-y}Te_y$ and $Ba_3Cu_{16-x}Se_{11-y}Te_y$ adopt the same rhombohedral structure with Cu_{26} clusters and CuQ_4 tetrahedra (Q = S, Se, Te). Cu_{26} clusters are interconnected through Cu–Cu contacts to form a three-dimensional network of Cu atoms, most of which have significant deficiencies (see Figure 3.6). Pure S (or Se), mixed Q and pure Te sites exist in this structure, resulting in a significant phase range for y.

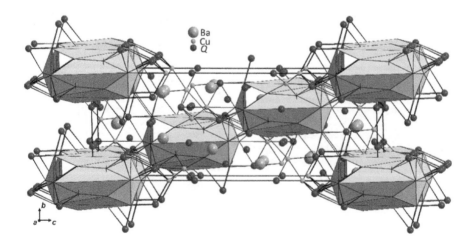

Figure 3.6. *Crystal structure of $Ba_3Cu_{16-x}Se_{11-y}Te_y$, emphasizing the Cu_{26} clusters*

Both the sulfide-telluride and the selenide-telluride are p-type semiconductors, according to the balanced formula $(Ba^{2+})_3(Cu^+)_{16}(Q^{2-})_{11}$, for x is always > 0. As expected, thermal conductivity values are low, which are consistently below 0.6 W m^{-1}K^{-1} for sulfide-telluride (Jafarzadeh et al. 2018) and below 0.7 W m^{-1}K^{-1} for selenide-telluride (Jafarzadeh et al. 2019). Repeat measurements showed that the sulfide-telluride remains stable under the measurement conditions, while the selenide-telluride does not. The instability of the selenide-telluride is caused by irreversible changes in the Cu site occupancies and can be inhibited by reducing the maximum temperature below 600 K. The best performing representative of these rhombohedral materials is $Ba_3Cu_{16-x}S_{11-y}Te_y$ with a competitive $zT_{max} = 0.88$ at 745 K.

$BaCu_{6-x}STe_6$ and $BaCu_{6-x}Se_{1-y}Te_{6+y}$ are isostructural, crystallizing in a relatively small cubic unit cell with distinct sites of the S/Se and Te positions, while some Te can replace Se in the latter case. The structure of $BaCu_{6-x}SeTe_6$ consists of Cu-deficient Cu cubes, centered by Se atoms and interconnected by Te_2^{2-} dumbbells (Figure 3.7). As with $A_2BaCu_8Te_{10}$, Ba atoms are located in pentagonal Cu_8Te_{12} dodecahedra.

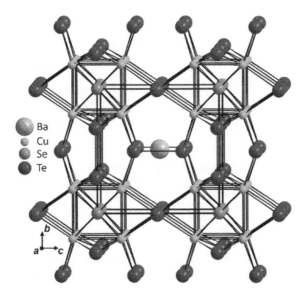

Figure 3.7. *Crystal structure of $BaCu_{6-x}SeTe_6$*

Charge balance may be hypothetically achieved according to $(Ba^{2+})(Cu^+)_6(Se^{2-})(Te_2^{2-})_3$ when x = 0. Despite the small unit cell, low conductivity values below 0.7 W m^{-1}K^{-1} are obtained because of the large Cu site defects, as less than 6 Cu positions per cube are occupied. In contrast to the above-mentioned examples, Cu atoms are not three-dimensionally connected. Cu cubes are separated from each other by Ba and Te atoms, which reduce Cu ion conductivity. Consequently, these materials do not experience stability problems caused by Cu ion migration. The best performance achieved so far was by BaCu$_{5.9}$SeTe$_6$ with zT_{max} = 0.81 at 600 K (Oudah et al. 2015). Higher temperatures cannot be tested as the material decomposes into binary and ternary chalcogenides above 650 K.

Figure 3.8 compares the leading barium copper chalcogenides. In addition to the common features of the *p*-type extrinsic semiconductor and low thermal conductivity, the figure of merit curves are very comparable. Finally, BaCu$_{2.04}$Te$_2$ has the highest zT values because its thermal stability is the highest in this group, enabling measurements up to 850 K.

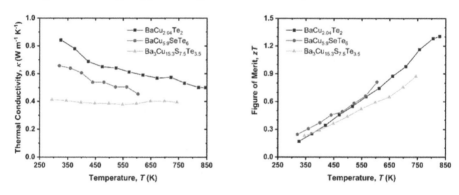

Figure 3.8. *Thermal conductivity (left) and figure of merit (right) of selected barium copper chalcogenides*

3.4. Conclusion

Investigations into the various copper and silver chalcogenides continue to result in further improvements, both with respect to their TE properties and their long-term stability.

This is of particular importance for TE energy conversion, because it restricts the use of the expensive, rare and toxic element tellurium, as many sulfides and selenides exhibit superior performance.

3.5. Acknowledgments

We thank the Natural Sciences and Engineering Research Council of Canada for financial support.

3.6. References

Assoud, A., Thomas, S., Sutherland, B., Zhang, H., Tritt, T.M., Kleinke, H. (2006). Thermoelectric properties of the new polytelluride $Ba_3Cu14-\delta Te12$. *Chemistry of Materials*, 18(16), 3866–3872.

Assoud, A., Cui, Y., Thomas, S., Sutherland, B., Kleinke, H. (2008). Structure and physical properties of the new telluride $BaAg_2Te_2$ and its quaternary variants $BaCu\delta Ag_2-\delta Te2$. *Journal of Solid State Chemistry*, 181(8), 2024–2030.

Balapanov, M.K., Gafurov, I.G., Mukhamed'yanov, U.K., Yakshibaev, R.A., Ishembetov, R.K. (2004). Ionic conductivity and chemical diffusion in superionic $Li_xCu2-xS$ ($0 \leq x \leq 0.25$). *Physica Status Solidi B*, 241(1), 114–119.

Ballikaya, S., Chi, H., Salvador, J.R., Uher, C. (2013). Thermoelectric properties of Ag-doped Cu_2Se and Cu_2Te. *Journal of Materials Chemistry A*, 1(40), 12478–12484.

Billeter, H. and Ruschewitz, U. (2008). Structural phase transitions in Ag_2Se (Naumannite). *Zeitschrift für Anorganische und Allgemeine Chemie*, 634(2), 241–246.

Biswas, K., He, J., Blum, I.D., Wu, C.-I., Hogan, T.P., Seidman, D.N., Dravid, V.P., Kanatzidis, M.G. (2012). High-performance bulk thermoelectrics with all-scale hierarchical architecture. *Nature*, 489(7416), 414–418.

Bourgès, C., Bouyrie, Y., Supka, A.R., Al Rahal Al Orabi, R., Lemoine, P., Lebedev, O.I., Ohta, M., Suekuni, K., Nassif, V., Hardy, V. et al. (2018). High-performance thermoelectric bulk colusite by process controlled structural disordering. *Journal of the American Chemical Society*, 140(6), 2186–2195.

Bouyrie, Y., Ohta, M., Suekuni, K., Kikuchi, Y., Jood, P., Yamamoto, A., Takabatake, T. (2017). Enhancement in the thermoelectric performance of colusites $Cu_{26}A_2E_6S_{32}$ (A = Nb, Ta; E = Sn, Ge) using E-site non-stoichiometry. *Journal of Materials Chemistry C*, 5(17), 4174–4184.

Brown, D.R., Day, T., Caillat, T., Snyder, G.J. (2013). Chemical stability of $(Ag,Cu)_2Se$: A historical overview. *Journal of Electronic Materials*, 42(7), 2014–2019.

Caillat, T., Fleurial, J.-P., Borshchevsky, A. (1997). Preparation and thermoelectric properties of semiconducting Zn4Sb3. *Journal of Physics and Chemistry of Solids*, 58(7), 1119–1125.

Dennler, G., Chmielowski, R., Jacob, S., Capet, F., Roussel, P., Zastrow, S., Nielsch, K., Opahle, I., Madsen, G.K.H. (2014). Are binary copper sulfides/selenides really new and promising thermoelectric materials? *Advanced Energy Materials*, 4(9), 1301581/1–12.

Ferhat, M. and Nagao, J. (2000). Thermoelectric and transport properties of β-Ag2Se compounds. *Journal of Applied Physics*, 88(2), 813–816.

Francioso, L., De Pascali, C., Farella, I., Martucci, C., Cretì, P., Siciliano, P., Perrone, A. (2011). Flexible thermoelectric generator for ambient assisted living wearable biometric sensors. *Journal of Power Sources*, 196, 3239–3243.

Freer, R. and Powell, A.V. (2020). Realising the potential of thermoelectric technology: A roadmap. *Journal of Materials Chemistry C*, 8(2), 441–463.

Fujikane, M., Kurosaki, K., Muta, H., Yamanaka, S. (2005). Thermoelectric properties of α- and β-Ag2Te. *Journal of Alloys and Compounds*, 393(1–2), 299–301.

Furlong, R.R. and Wahlquist, E.J. (1999). US space missions using radioisotope power systems. *Nuclear News*, 42, 26–35.

Gahtori, B., Bathula, S., Tyagi, K., Jayasimhadri, M., Srivastava, A.K., Singh, S., Budhani, R.C., Dhar, A. (2015). Giant enhancement in thermoelectric performance of copper selenide by incorporation of different nanoscale dimensional defect features. *Nano Energy*, 13, 36–46.

Guo, Q., Chan, M., Kuropatwa, B.A., Kleinke, H. (2013). Enhanced thermoelectric properties of variants of Tl9SbTe6 and Tl9BiTe6. *Chemistry of Materials*, 25(20), 4097–4104.

Guo, Q., Assoud, A., Kleinke, H. (2014). Improved bulk materials with thermoelectric figure-of-merit > 1: Tl10–xSnxTe6 and Tl10–xPbxTe6. *Adv. Energy Mater.*, 4(14), 1400348/1–8.

Guo, K., Lin, J., Li, Y., Zhu, Y., Li, X., Yang, X., Xing, J., Yang, J., Luo, J., Zhao, J.-T. (2020). Suppressing the dynamic precipitation and lowering the thermal conductivity for stable and high thermoelectric performance in BaCu2Te2 based materials. *Journal of Materials Chemistry A*, 8(10), 5323–5331.

Harman, T.C., Walsh, M.P., Laforge, B.E., Turner, G.W. (2005). Nanostructured thermoelectric materials. *J. Electron. Mater.*, 34(5), L19–L22.

He, J. and Tritt, T.M. (2017). Advances in thermoelectric materials research: Looking back and moving forward. *Science (New York, N.Y.)*, 357(6358), eaak9997/1-9.

He, Y., Day, T., Zhang, T., Liu, H., Shi, X., Chen, L., Snyder, G.J. (2014). High thermoelectric performance in non-toxic earth-abundant copper sulfide. *Advanced Materials*, 26(23), 3974–3978.

He, Y., Zhang, T., Shi, X., Wei, S.-H., Chen, L. (2015). High thermoelectric performance in copper telluride. *NPG Asia Materials*, 7(8), e210/1-7.

Heo, J., Laurita, G., Muir, S., Subramanian, M.A., Keszler, D.A. (2014). Enhanced thermoelectric performance of synthetic tetrahedrites. *Chemistry of Materials*, 26(6), 2047–2051.

Hu, Q., Zhu, Z., Zhang, Y., Li, X.-J., Song, H. (2018). Remarkably high thermoelectric performance of $Cu_{2-x}Li_xSe$ bulks with nanopores. *Journal of Materials Chemistry A*, 6(46), 23417–23424.

Hu, Q., Zhang, Y., Zhang, Y., Li, X.J., Song, H. (2020). High thermoelectric performance in Cu2Se/CDs hybrid materials. *Journal of Alloys and Compounds*, 813, 152204.

Huang, S., Wei, T.-R., Chen, H., Xiao, J., Zhu, M., Zhao, K., Shi, X. (2021). Thermoelectric Ag2Se: Imperfection, homogeneity, and reproducibility. *ACS Applied Materials & Interfaces*, 13(50), 60192–60199.

Jafarzadeh, P., Oudah, M., Assoud, A., Farahi, N., Müller, E., Kleinke, H. (2018). High thermoelectric performance of $Ba_3Cu_{16-x}(S,Te)_{11}$. *Journal of Materials Chemistry C*, 6(47), 13043–13048.

Jafarzadeh, P., Assoud, A., Ramirez, D., Farahi, N., Zou, T., Müller, E., Kycia, J.B., Kleinke, H. (2019). Thermoelectric properties and stability of $Ba_3Cu_{16-x}Se_{11-y}Te_y$. *Journal of Applied Physics*, 126(2), 25109/1–9.

Kleinke, H. (2010). New bulk materials for thermoelectric power generation: Clathrates and complex antimonides. *Chemistry of Materials*, 22(3), 604–611.

Kuropatwa, B.A., Cui, Y., Assoud, A., Kleinke, H. (2009). Crystal structure and physical properties of the new selenide–tellurides $Ba_3Cu_{17-x}(Se,Te)_{11}$. *Chemistry of Materials*, 21(1), 88–93.

Kuropatwa, B.A., Assoud, A., Kleinke, H. (2011). Crystal structure and physical properties of the new chalcogenides $Ba_3Cu_{17-x}(S,Te)_{11}$ and $Ba_3Cu_{17-x}(S,Te)_{11.5}$ with two different cu clusters. *Inorganic Chemistry*, 50(16), 7831–7837.

Kurosaki, K., Kosuga, A., Muta, H., Uno, M., Yamanaka, S. (2005a). Ag9TlTe5: A high-performance thermoelectric bulk material with extremely low thermal conductivity. *Applied Physics Letters*, 87(6), 061919/1–3.

Kurosaki, K., Uneda, H., Muta, H., Yamanaka, S. (2005b). Thermoelectric properties of potassium-doped β-BaCu2S2 with natural superlattice structure. *Journal of Applied Physics*, 97(5), 053705/1–4.

Lee, Y.K., Luo, Z., Cho, S.P., Kanatzidis, M.G., Chung, I. (2019). Surface oxide removal for polycrystalline SnSe reveals near-single-crystal thermoelectric performance. *Joule*, 3(3), 719–731.

Leonov, V. and Vullers, R.J.M. (2009). Wearable electronics self-powered by using human body heat: The state of the art and the perspective. *Journal of Renewable and Sustainable Energy*, 1, 062701/1–17.

Li, J., Zhao, L.-D., Sui, J., Berardan, D., Cai, W., Dragoe, N. (2015). BaCu2Se2 based compounds as promising thermoelectric materials. *Dalton Trans.*, 44(5), 2285–2293.

Li, M., Cortie, D.L., Liu, J., Yu, D., Islam, S.M.K.N., Zhao, L., Mitchell, D.R.G., Mole, R.A., Cortie, M.B., Dou, S. et al. (2018). Ultra-high thermoelectric performance in graphene incorporated Cu2Se: Role of mismatching phonon modes. *Nano Energy*, 53, 993–1002.

Liu, H., Shi, X., Xu, F., Zhang, L., Zhang, W., Chen, L., Li, Q., Uher, C., Day, T., Snyder, G.J. (2012). Copper ion liquid-like thermoelectrics. *Nature Materials*, 11(5), 422–425.

Long, S.O.J., Powell, A.V., Vaqueiro, P., Hull, S. (2018). High thermoelectric performance of bornite through control of the Cu(II) content and vacancy concentration. *Chemistry of Materials*, 30(2), 456–464.

Lv, H.Y., Liu, H.J., Shi, J., Tang, X.F., Uher, C. (2013). Optimized thermoelectric performance of Bi2Te3 nanowires. *Journal of Materials Chemistry A*, 1, 6831–6838.

Mao, T., Qiu, P., Du, X., Hu, P., Zhao, K., Xiao, J., Shi, X., Chen, L. (2020). Enhanced thermoelectric performance and service stability of Cu2Se via tailoring chemical compositions at multiple atomic positions. *Advanced Functional Materials*, 30(6), 1908315/1–8.

Matsumoto, M., Mori, M., Haraguchi, T., Ohtani, M., Kubo, T., Matsumoto, K., Matsuda, H. (2015). Development of state of the art compact and lightweight thermoelectric generator using vacuum space structure. *SAE International Journal of Engines*, 8, 1815–1825.

Mayasree, O., Cui, Y., Assoud, A., Kleinke, H. (2010). Structure change via partial Se/Te substitution: Crystal structure and physical properties of the telluride Ba2Cu4−xTe5 in contrast to the selenide-telluride Ba2Cu4−xSeyTe5−y. *Inorg. Chem.*, 49(14), 6518–6524.

Mayasree, O., Sankar, C.R., Cui, Y., Assoud, A., Kleinke, H. (2011). Synthesis, structure, and thermoelectric properties of barium copper polychalcogenides with chalcogen-centered Cu clusters and Te22-dumbbells. *European Journal of Inorganic Chemistry*, 2011, 4037–4042.

Mayasree, O., Sankar, C.R., Kleinke, K.M., Kleinke, H. (2012). Cu clusters and chalcogen–chalcogen bonds in various copper polychalcogenides. *Coordination Chemistry Reviews*, 256(13–14), 1377–1383.

Nunna, R., Qiu, P., Yin, M., Chen, H., Hanus, R., Song, Q., Zhang, T., Chou, M.-Y., Agne, M.T., He, J. et al. (2017). Ultrahigh thermoelectric performance in Cu2Se-based hybrid materials with highly dispersed molecular CNTs. *Energy & Environmental Science*, 10(9), 1928–1935.

Olvera, A.A., Moroz, N.A., Sahoo, P., Ren, P., Bailey, T.P., Page, A.A., Uher, C., Poudeu, P.F.P. (2017). Partial indium solubility induces chemical stability and colossal thermoelectric figure of merit in Cu_2Se. *Energy & Environmental Science*, 10(7), 1668–1676.

Orr, B., Akbarzadeh, A., Mochizuki, M., Singh, R. (2016). A review of car waste heat recovery systems utilising thermoelectric generators and heat pipes. *Applied Thermal Engineering*, 101, 490–495.

Oudah, M., Kleinke, K.M., Kleinke, H. (2015). Thermoelectric properties of the quaternary chalcogenides BaCu5.9STe6 and BaCu5.9SeTe6. *Inorganic Chemistry*, 54(3), 845–849.

Paccard, D., Paccard, L., Brun, G., Tedenac, J.C. (1992). A new phase in the Tl–Ag–Te system: Crystal structure of Tl2Ag16Te11. *Journal of Alloys and Compounds*, 184, 337–342.

Park, H., Lee, D., Park, G., Park, S., Khan, S., Kim, J., Kim, W. (2019). Energy harvesting using thermoelectricity for IoT (Internet of Things) and E-skin sensors. *Journal of Physics: Energy*, 1(4), 42001/1–15.

Patschke, R., Zhang, X., Singh, D., Schindler, J., Kannewurf, C.R., Lowhorn, N., Tritt, T., Nolas, G.S., Kanatzidis, M.G. (2001). Thermoelectric properties and electronic structure of the cage compounds A2BaCu8Te10 (A = K, Rb, Cs): Systems with low thermal conductivity. *Chem. Mater.*, 13(2), 613–621.

Pavan Kumar, V., Guélou, G., Lemoine, P., Raveau, B., Supka, A.R., Al Rahal Al Orabi, R., Fornari, M., Suekuni, K., Guilmeau, E. (2019). Copper-rich thermoelectric sulfides: Size-mismatch effect and chemical disorder in the [TS4]Cu6 complexes of Cu26T2Ge6S32 (T = Cr, Mo, W) Colusites. *Angewandte Chemie International Edition*, 131(43), 15601–15609.

Pei, Y., LaLonde, A., Iwanaga, S., Snyder, G.J. (2011). High thermoelectric figure of merit in heavy hole dominated PbTe. *Energy and Environmental Science*, 4(6), 2085–2089.

Qiu, P., Zhang, T., Qiu, Y., Shi, X., Chen, L. (2014). Sulfide bornite thermoelectric material: A natural mineral with ultralow thermal conductivity. *Energy and Environmental Science*, 7(12), 4000–4006.

Qiu, P., Shi, X., Chen, L. (2016). Cu-based thermoelectric materials. *Energy Storage Materials*, 3, 85–97.

Sharma, R.C. and Chang, Y.A. (1986). The Ag−S (Silver–Sulfur) system. *Bulletin of Alloy Phase Diagrams*, 7(3), 263–269.

Shi, X., Yang, J., Salvador, J.R., Chi, M., Cho, J.Y., Wang, H., Bai, S., Yang, J., Zhang, W., Chen, L. (2011). Multiple-filled skutterudites: High thermoelectric figure of merit through separately optimizing electrical and thermal transports. *Journal of the American Chemical Society*, 133(20), 7837–7846.

Shi, Y., Assoud, A., Ponou, S., Lidin, S., Kleinke, H. (2018). A new material with a composite crystal structure causing ultralow thermal conductivity and outstanding thermoelectric properties: Tl2Ag12Te7+δ. *Journal of the American Chemical Society*, 140(27), 8578–8585.

Shi, Y., Sturm, C., Kleinke, H. (2019). Chalcogenides as thermoelectric materials. *Journal of Solid State Chemistry*, 270, 273–279.

Shi, X.L., Zou, J., Chen, Z.G. (2020). Advanced thermoelectric design: From materials and structures to devices. *Chemical Reviews*, 120(15), 7399–7515.

Snyder, G.J. and Toberer, E.S. (2008). Complex thermoelectric materials. *Nature Materials*, 7(2), 105–114.

Sturm, C., Boccalon, N., Assoud, A., Zou, T., Kycia, J., Kleinke, H. (2021). Thermoelectric properties of hot-pressed Ba3Cu14−δTe12. *Inorganic Chemistry*, 60(17), 12781–12789.

Su, X., Fu, F., Yan, Y., Zheng, G., Liang, T., Zhang, Q., Cheng, X., Yang, D., Chi, H., Tang, X. et al. (2014). Self-propagating high-temperature synthesis for compound thermoelectrics and new criterion for combustion processing. *Nature Communications*, 5, 4908/1–7.

Tafti, M.Y., Ballikaya, S., Khachatourian, A.M., Noroozi, M., Saleemi, M., Han, L., Nong, N.V., Bailey, T., Uher, C., Toprak, M.S. (2016). Promising bulk nanostructured Cu2Se thermoelectrics via high throughput and rapid chemical synthesis. *RSC Advances*, 6(112), 111457–111464.

Tang, J., Qin, C., Yu, H., Zeng, Z., Cheng, L., Ge, B., Chen, Y., Li, W., Pei, Y. (2022). Ultralow lattice thermal conductivity enables high thermoelectric performance in BaAg2Te2 alloys. *Materials Today Physics*, 22, 100591/1–8.

Tian, R., Liu, Y., Koumoto, K., Chen, J. (2019). Body heat powers future electronic skins. *Joule*, 3(6), 1399–1403.

Toberer, E.S., Cox, C.A., Brown, S.R., Ikeda, T., May, A.F., Kauzlarich, S.M., Snyder, G.J. (2008). Traversing the metal-insulator transition in a Zintl phase: Rational enhancement of thermoelectric efficiency in Yb14Mn1-xAlxSb11. *Advanced Functional Materials*, 18, 2795–2800.

Venkatasubramanian, R., Slivola, E., Colpitts, T., O'Quinn, B. (2001). Thin-film thermoelectric devices with high room-temperature figures of merit. *Nature*, 413(6856), 597–602.

Wang, Y.C. and DiSalvo, F.J. (2001). Structure and physical properties of BaCu2Te2. *Journal of Solid State Chemistry*, 156(1), 44–50.

Wang, T., Chen, H.Y., Qiu, P.F., Shi, X., Chen, L.D. (2019). Thermoelectric properties of Ag2S superionic conductor with intrinsically low lattice thermal conductivity. *Acta Physica Sinica*, 68(9), 090201/1–9.

Wei, T., Qin, Y., Deng, T., Song, Q., Jiang, B., Liu, R., Qiu, P., Shi, X., Chen, L. (2019). Copper chalcogenide thermoelectric materials. *Science China Materials*, 62(1), 8–24.

Yang, J. and Caillat, T. (2006). Thermoelectric materials for space and automotive power generation. *MRS Bulletin*, 31(03), 224–229.

Yang, J. and Stabler, F.R. (2009). Automotive applications of thermoelectric materials. *Journal of Electronic Materials*, 38(7), 1245–1251.

Yang, Y., Wei, X.-J., Liu, J. (2007). Suitability of thermoelectric power generator for implantable medical devices. *Journal of Physics D: Applied Physics*, 5790–5800.

Yang, D., Su, X., Meng, F., Wang, S., Yan, Y., Yang, J., He, J., Zhang, Q., Uher, C., Kanatzidis, M.G. et al. (2017). Facile room temperature solventless synthesis of high thermoelectric performance Ag2Se via a dissociative adsorption reaction. *Journal of Materials Chemistry A*, 5(44), 23243–23251.

Yang, C., Guo, K., Yang, X., Xing, J., Wang, K., Luo, J., Zhao, J.-T. (2019). Realizing high thermoelectric performance in BaCu2–xAgxTe2 through enhanced carrier effective mass and point-defect scattering. *ACS Applied Energy Materials*, 2(1), 889–895.

Zaia, E.W., Gordon, M.P., Yuan, P., Urban, J.J. (2019). Progress and perspective: Soft thermoelectric materials for wearable and internet-of-things applications. *Advanced Electronic Materials*, 5(11), 1800823/1–20.

Zhao, L., Wang, X., Fei, F.Y., Wang, J., Cheng, Z., Dou, S., Wang, J., Snyder, G.J. (2015a). High thermoelectric and mechanical performance in highly dense $Cu_{2-x}S$ bulks prepared by a melt-solidification technique. *Journal of Materials Chemistry A*, 3(18), 9432–9437.

Zhao, L., Wang, X., Wang, J., Cheng, Z., Dou, S., Wang, J., Liu, L. (2015b). Superior intrinsic thermoelectric performance with zT of 1.8 in single-crystal and melt-quenched highly dense $Cu_{2-x}Se$ bulks. *Scientific Reports*, 5(1), 7671/1–6.

Zhao, S., Chen, H., Zhao, X., Luo, J., Tang, Z., Zeng, G., Yang, K., Wei, Z., Wen, W., Chen, X. et al. (2020). Excessive iodine addition leads to room-temperature superionic Cu_2S with enhanced thermoelectric properties and improved thermal stability. *Materials Today Physics*, 15, 100271/1–8.

Zhou, C., Lee, Y.K., Yu, Y., Byun, S., Luo, Z.Z., Lee, H., Ge, B., Lee, Y.L., Chen, X., Lee, J.Y. et al. (2021). Polycrystalline SnSe with a thermoelectric figure of merit greater than the single crystal. *Nature Materials*, 20(10), 1378–1384.

Zhu, Z., Zhang, Y., Song, H., Li, X.-J. (2019). High thermoelectric performance and low thermal conductivity in $Cu_{2-x}Na_xSe$ bulk materials with micro-pores. *Applied Physics A*, 125(8), 572/1–7.

Zhu, T., Bai, H., Zhang, J., Tan, G., Yan, Y., Liu, W., Su, X., Wu, J., Zhang, Q., Tang, X. (2020). Realizing high thermoelectric performance in Sb-Doped Ag_2Te compounds with a low-temperature monoclinic structure. *ACS Applied Materials & Interfaces*, 12(35), 39425–39433.

4
Sulfide Thermoelectrics: Materials and Modules

Michihiro OHTA, Priyanka JOOD and Kazuki IMASATO
Global Zero Emission Research Center, National Institute of Advanced Industrial Science and Technology (AIST), Japan

4.1. Introduction

A large number of materials that offer various benefits such as high-performance, high-temperature operation and high mechanical strength have been developed for thermoelectric applications. Among them, sulfides are promising candidates for environmental-friendly and cost-effective thermoelectrics due to the following reasons: firstly, sulfur is an earth-abundant element; secondly, the electronegativity of sulfur is lower in the oxygen family (oxygen, sulfur, selenium and tellurium); thus, the electrical resistivity (ρ) of sulfides are generally low; thirdly, a reduction in the lattice thermal conductivity (κ_l) for sulfides has been demonstrated through various strategies to suppress phonon propagation, despite sulfur's low atomic mass. For example, complex crystal structure, defects and disorder, and liquid-like behavior in sulfide result in the low lattice thermal conductivity. The low ρ (second reason) and low κ_l (third reason) result in a high thermoelectric figure of merit (zT).

Firstly, this chapter discusses the recent progress in thermoelectric sulfides, including rare-earth sulfides, layered sulfides, Pb–Bi–S-based systems, Cu and Ag sulfides-based superionic conductors, Cu–S-based tetrahedrites and colusites, Chevrel-phase sulfides and chalcopyrites. Secondly, this chapter addresses recent attempts to develop thermoelectric modules based on these sulfides for waste heat recovery. Finally, the chapter concludes with a brief overview of future opportunities and potential challenges in the field of sulfide thermoelectric performance enhancement and application. This chapter is recommended to be studied in conjunction with the previously published review articles to gain a better understanding of thermoelectric sulfides (Jood and Ohta 2015; Suekuni and Takabatake 2016; Hébert et al. 2016; Powell 2019; Guélou et al. 2021).

4.2. Materials

4.2.1. *Rare-earth sulfides*

Rare-earth sesqui-chalcogenides, Ln_2Ch_3 (Ln: rare-earth element, Ch: chalcogen), with a cubic Th_3P_4-type structure (γ-phase) are promising n-type high-temperature thermoelectric materials studied for more than half a century due to their high melting point, defect structure and self-doping ability (Cutler et al. 1964; Gschneidner et al. 1987; Wood 1988; Gschneidner 1998). This section mainly focuses on the preparation and high-temperature thermoelectric properties of γ-phase in rare-earth sulfides γ-Ln_2S_3 among rare-earth chalcogenides. Furthermore, γ-Ln_2S_3 has a high melting point; for example, the melting point of γ-$Pr_{2+x}S_3$ is about 2,300 K (Gschneidner 1992). Therefore, this system is suited for high-temperature applications due to its high zT at a high-temperature and melting point.

Various methods are used to synthesize rare-earth sulfides, including the direct reaction of the elemental components and sulfurization of the rare-earth oxides/salts with C_2S or H_2S gases (Gschneidner et al. 1987; Gschneidner 1998; Henderson et al. 1967; Toide et al. 1973; Guittard and Flahaut 1991; Hirai et al. 1998, 2003; Ohta et al. 2004, 2009a, 2009b; Yuan et al. 2010). The physical properties of rare-earth sulfides with different chemical compositions can be studied using a direct reaction of the components that can be precisely controlled (Gschneidner et al. 1987; Gschneidner 1998). Alternatively, the CS_2 or H_2S sulfurization processes could be employed for industrial-scale production. The value of the standard free-energy change, ΔG, for CS_2 sulfurization of rare-earth oxides is lower than that for H_2S sulfurization of rare-earth oxides over the temperature range of 600–1,300 K (Henderson et al. 1967; Hirai et al. 1998, 2003; Ohta et al. 2004). This implies that CS_2 gas is a powerful sulfurizing agent for rare-earth oxides, allowing the low-temperature formation of rare-earth sulfides.

The γ-Ln_2S_3 compounds contain vacancies randomly distributed over cation sites (Flahaut et al. 1979). The self-doping ability arises because the vacancies can be filled with additional Ln atoms. Therefore, the γ-phases are formed over a compositional range between $Ln_2V_{0.25}S_3$ (V: vacancy) and $Ln_{2.25}S_3$. Because one Ln^{3+} ion contributes three electrons and one S^{2-} ion accepts two electrons, Ln self-doping allows for the tuning of the carrier concentration, improving the thermoelectric power factor. Figure 4.1(a)–(c) shows the electrical resistivity, the Seebeck coefficient and power factor versus chemical x in $Ln_{2+x}S_3$ ($0 \leq x \leq 0.25$) (Takeshita et al. 1985; Wood et al. 1985; Gadzhiev et al. 2000; Ohta et al. 2008, 2011; Ohta and Hirai 2009). The Seebeck coefficient (see Figure 4.1(a)) and electrical resistivity (see Figure 4.1(b)) decrease as x increases. The electrical resistivity and the Seebeck coefficient in $La_{2+x}S_3$ decrease from ~100 $\mu\Omega$ m and ~−230 μW K^{-1} to ~10 $\mu\Omega$ m and ~−95 μW K^{-1}, respectively, at 1,400 K as Ln content increases from 0.03 to 0.22. The power factor is improved by changing the Ln content (see Figure 4.1(c)). A power factor of ~970 μW m^{-1} K^2 is obtained for $La_{2.17}S_3$ at 1,400 K. The carrier concentration can also be tuned by occupying the vacancies with divalent rare-earth (e.g. Sm, Eu and Yb) and alkaline earth (e.g. Ca and Ba) metals, which improves the power factor (Nakahara et al. 1988; Katsuyama et al. 1997).

Moreover, the numerous randomly distributed Ln vacancies result in effective phonon scattering, yielding total and low lattice thermal conductivity (Snyder and Toberer 2008). Figure 4.1(d) shows the temperature dependence of total thermal conductivity (Wood et al. 1985; Gadzhiev et al. 2000; Ohta and Hirai 2009; Ohta et al. 2011). The total thermal conductivity for all samples is less than 4.0 W m^{-1} K^{-1}. The lattice thermal conductivity, κ_l, in $La_{2.03}S_3$ at 1,400 K, which was calculated by subtracting the electronic thermal conductivity, κ_e, from the total thermal conductivity, κ ($\kappa_l = \kappa - \kappa_e$), is 0.83 W m^{-1} K^{-1}, where the carrier thermal conductivity was estimated from the electrical resistivity, ρ, using the Wiedemann–Franz law with the Lorenz number, L, of 2.45×10^{-8} W Ω K^{-2} ($\kappa_e = LT/\rho$).

Figure 4.1(e) shows the temperature dependence of zT. The zT value was optimized via chemical composition tuning. In $Gd_{2+x}S_3$ at 1,270 K, the zT value has a maximum ($zT \sim 0.8$) at $x = 0.03$.

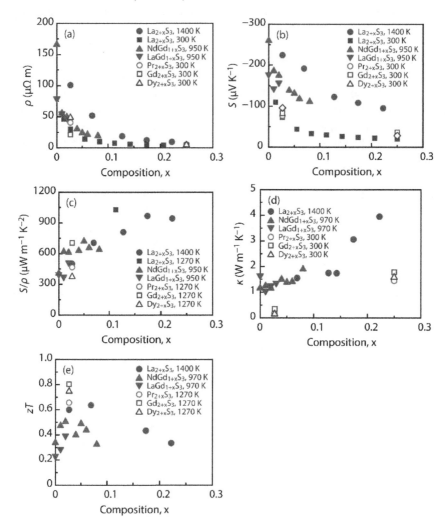

Figure 4.1. (a) Electrical resistivity, ρ; (b) the Seebeck coefficient, S; (c) power factor, S^2/ρ; (d) total thermal conductivity, κ and (e) thermoelectric figure of merit, zT, versus chemical composition, x ($0 \leq x \leq 0.25$), in $La_{2+x}S_3$ (Wood et al. 1985), $La_{2+x}S_3$ (Takeshita et al. 1985), $NdGd_{1+x}S_3$ (Ohta and Hirai 2009), $LaGd_{1+x}S_3$ (Ohta et al. 2011), $Pr_{2+x}S_3$ (Gadzhiev et al. 2000), $Gd_{2+x}S_3$ (Gadzhiev et al. 2000) and $Dy_{2+x}S_3$ (Gadzhiev et al. 2000). For a color version of this figure, see www.iste.co.uk/akinaga/thermoelectric1.zip

4.2.2. Layered sulfides

The two-dimensional crystal structure of layered systems such as TiS_2 and misfit-layered sulfides $[MS]_{1+m}[TS_2]_n$ (where M = Sn, Pb, Sb, Bi and rare-earth metals; T = Ti, V, Cr, Nb and Ta; n = 1, 2, 3) provides an ideal platform for simultaneously regulating electrical and thermal transports (Jood and Ohta 2015; Koyano et al. 1986; Wiegers 1996; Logothetis et al. 1980). These materials are a perfect example of the renowned phonon glass-electron crystal behavior (PGEC) proposed by Slack (1995). The research on layered sulfides gained traction after Han and Cook (1994) discovered a high power factor of ~1,500 μW m^{-1} K^{-2} at 300 K for $Ti_{1+x}S_2$ alloy, and Imai et al. (2001) later reported a high power factor of ~3,710 μW m^{-1} K^{-2} at 300 K in a near-stoichiometric TiS_2 single crystal (see Table 4.1). However, the zT ~0.16 at 300 K in the in-plane direction remained low due to the high lattice thermal conductivity of ~6.4 W m^{-1} K^{-1}.

Most research was then focused on decreasing the polycrystalline TiS_2 system's lattice thermal conductivity, while simultaneously tuning the carrier concentration by intercalating guest atoms (such as Ti, Cu, Co and Ag) (Guilmeau et al. 2011; Ohta et al. 2012; Beaumale et al. 2014; Barbier et al. 2015; Guélou et al. 2016; Guilmeau et al. 2017) or rock salt type MS layers (where M = Pb, Bi and Sn) (Wan et al. 2010, 2011, 2012; Putri et al. 2013; Yin et al. 2018) into the van der Waals gap of the host TiS_6 octahedral layers, pushing the zT close to ~0.5 at 823 K (for $Cu_{0.1}TiS_2$ in the in-plane direction (see Table 4.1) (Guilmeau et al. 2011, 2017)). The host TiS_2 layer in the intercalated TiS_2 provides a high Seebeck coefficient, and the electron pathway (i.e. low electrical resistivity) and MS layer reduces the lattice thermal conductivity due to increased lattice disorder. The interfaces between the layers scatter both the charge carriers and heat-carrying phonons, making these layered systems have anisotropic transport properties. Texture engineering in TiS_2 has been recently reported to increase the zT to ~0.7 at 673 K (zT ~ 0.15 at 300 K) in the in-plane direction due to the 20% reduced lattice thermal conductivity arising from intensified grain boundary phonon scattering (see Table 4.1) (Gu et al. 2020). Alternatively, ionized impurity scattering was introduced by forming TiS_2–$AgSnSe_2$ composites, enabling a high power factor (~ 1,550 μW m^{-1} K^{-2}) at 700 K while achieving low lattice thermal conductivity (~1.0 W m^{-1} K^{-1} at 700 K). A high zT of ~0.8 at 700 K was reported for TiS$_2$–4%AgSnSe$_2$ composite (see Table 4.1) (Wang et al. 2018). Flexible hybrid organic/TiS$_2$ superlattices have been fabricated possessing an ultralow lattice thermal conductivity of ~0.12 W m^{-1} K^{-1} and a promising zT of ~0.28 at 373 K, creating the opportunity for layered sulfides to be used in flexible thermoelectric devices (see Table 4.1) (Wan et al. 2015; Tian et al. 2017).

The other family of misfit-layered sulfides ([MS]$_{1+m}$[TS$_2$]$_n$, T = V, Cr, Nb and Ta), unlike TiS$_2$-based layered sulfides, has received little attention for their thermoelectric properties and only a few reports exist so far on ([MS]$_{1+m}$TS$_2$, M = lanthanide series, T = Cr, Nb) (Miyazaki et al. 2004, 2013; Jood et al. 2014, 2015; Sotnikov et al. 2020). However, their crystal structure and physical properties were extensively studied in the 1990s (Wiegers 1996). Among all compounds in the LnS-based series, [Yb$_{1.90}$S$_2$]$_{0.62}$NbS$_2$ shows the highest power factor (~250 μW m^{-1} K^{-2}) and zT (~0.11) at 300 K due to the low carrier concentration (high Seebeck coefficient) attributable to Yb deficiency from the distorted atomic arrangement (see Table 4.1) (Miyazaki et al. 2004, 2013).

Misfit compounds are highly sensitive to microstructural engineering because of the anisotropic nature of their atomic bonds. Therefore, zT can be enhanced by tuning the microstructure, as was achieved for (LaS)$_{1+m}$TS$_2$ (T = Nb, Cr) by varying the sulfurization duration from 6 to 12 h, followed by pressure-assisted sintering to obtain samples with randomly and highly oriented textures (Jood et al. 2014). The highly oriented texture produced the highest in-plane zT of ~0.14 at 950 K among the (LaS)$_{1.20}$CrS$_2$ systems, whereas the weakly/randomly oriented texture produced the highest in-plane zT of ~0.15 at 950 K among the (LaS)$_{1.14}$NbS$_2$ systems. Disorder and defects in the LaS layer, such as strain-induced stacking disorder, were observed in both CrS$_2$- and NbS$_2$-based compounds (Jood et al. 2014, 2015). These defects could be further tuned via chemical composition tuning. For example, the best structural and long-range ordering was found in the La rich sample, (La$_{1.05}$S$_{1.05}$)$_{1.14}$NbS$_2$, which resulted in a highly textured grain growth in-plane, pushing the in-plane zT to about ~0.2 at 950 K (see Table 4.1) (Jood et al. 2015).

Sample	Direction and T	ρ	S	S^2/ρ	κ_l	zT	References
Near-stoichiometric TiS$_2$ (single crystal)	In-plane 300 K	17	−250	3,710	6.4	0.16	Imai et al. (2001)
	Out-of-plane 300 K	13,000	-	-	4.2	-	Imai et al. (2001)
Ti$_{1.008}$S$_2$	In-plane 660 K	17.4	−150	1,270	1.8	0.34	Ohta et al. (2012)
	Out-of-plane 660 K	33.0	−150	730	1.8	0.21	Ohta et al. (2012)
Ti$_{1.025}$S$_2$	In-plane 700 K	34	−190	1,080	1.1	0.48	Beaumale et al. (2014)
Textured TiS$_2$	In-plane 323 K	7	−110	1,670	2.5	0.16 (0.7 at 673 K)	Gu et al. (2020)
Cu$_{0.1}$TiS$_2$	In-plane 800 K	19	−140	1,060	0.8	0.47	Guilmeau et al. (2011)
(BiS)$_{1.2}$(TiS$_2$)$_2$	In-plane 770 K	10	−90	810	0.3	0.27	Wan et al. (2010, 2012)
(SnS)$_{1.2}$(TiS$_2$)$_2$	In-plane 670 K	20	−140	900	0.9	0.35	Wan et al. (2011, 2012)
(SnS)$_{1.2}$(Cu$_{0.02}$Ti$_{0.98}$S$_2$)$_2$	In-plane 720 K	30	−140	670	0.6	0.42	Yin et al. (2018)
TiS$_2$–4% AgSnSe$_2$ composite	In-plane 700 K	34	−230	1,550	1.0	0.8	Wang et al. (2018)
TiS$_2$[(HA)$_{0.08}$(H$_2$O)$_{0.22}$ (DMSO)$_{0.03}$]	In-plane 373 K	20	−90	400	0.12 (300 K)	0.28	Wan et al. (2015)
(Yb$_{1.90}$S$_2$)$_{0.62}$NbS$_2$	In-plane 300 K	14	60	250	0.8	0.1	Miyazaki et al. (2004)
(La$_{1.05}$S$_{1.05}$)$_{1.14}$NbS$_2$	In-plane 950 K	16	90	460	0.9	0.20	Jood et al. (2015)

Table 4.1. *High-temperature Seebeck coefficient, S ($\mu V\ K^{-1}$), electrical resistivity, ρ ($\mu\Omega\ m$), power factor, S^2/ρ ($\mu W\ m^{-1}\ K^{-2}$), lattice thermal conductivity, κ_l ($W\ m^{-1}\ K^{-1}$) and thermoelectric figure of merit, zT, of TiS$_2$-based, CrS$_2$-based and NbS$_2$-based systems*

4.2.3. Pb–Bi–S-based systems

Bismuth sulfide (Bi$_2$S$_3$) and lead sulfide (PbS) are known as the family of well-studied/high-performance Bi$_2$Te$_3$ and PbTe. PbS has the same rock salt structure as PbTe and a slightly higher bandgap, E_g, of ~0.41 eV (PbTe: E_g ~ 0.32 eV) (Scanlon et al. 1959). Although PbS shows a slightly lower power factor and higher lattice thermal conductivity than PbTe, it can be useful in avoiding the use of Te, which is in short supply. The terrestrial abundance of sulfur (3.5×10^5 ppb) significantly exceeds that of tellurium (1 ppb). Moreover, PbS has been studied for higher temperature applications due to the large E_g and high melting point, T_m, of PbS ($T_m \sim 1{,}391$ K for PbS, $T_m \sim 1{,}197$ K for PbTe). The thermoelectric performance was enhanced by nanostructuring and substitutions/interstitials (Johnsen et al. 2011; Zhao et al. 2011; Qin et al. 2021); zT of ~1.2 was obtained in $Pb_{0.99}Cu_{0.01}S_{-0.01}Cu$ at 773 K (Qin et al. 2021). Because PbS has the same crystal structure as PbTe/PbSe, alloying with these compounds (Pb(S, Se, Te) alloys) boosts the thermoelectric performance (Girard et al. 2011; Korkosz et al. 2014; Wu et al. 2014).

Bi_2S_3 has also been investigated from the perspective of environmental friendliness and cost-effectiveness of sulfides. Although Bi_2Te_3 shows a high thermoelectric performance among all thermoelectric materials, it has some limitations in the operating temperature. Similar to the renowned Bi_2Te_3, Bi_2S_3 possesses low lattice thermal conductivity (~0.5 W m^{-1} K^{-1} at 500 K) based on its complex crystal structure (Biswas et al. 2012). However, the thermoelectric performance is limited by the relatively low intrinsic carrier concentration (<10^{18} cm^{-3}) and large bandgap ($E_g \sim 1.3$ eV) due to the strong bonding between Bi and S. Manipulating the carrier concentration was achieved in several investigations by controlling sulfur vacancies (Mizoguchi et al. 1995) or using metallic and halogen dopants (Guo et al. 2020a, 2020b). These efforts result in zT of about 0.8 at 760 K with the micro/nanostructured polycrystalline Bi_2S_3 (Ji et al. 2021).

The Pb–Bi–S-based systems – such as $PbBi_2S_4$, $Pb_3Bi_2S_6$ and $CdPb_2Bi_4S_9$ (Ohta et al. 2014; Zhao et al. 2019a; Cai et al. 2021) – have recently received attention. $PbBi_2S_4$, $Pb_3Bi_2S_6$ and $CdPb_2Bi_4S_9$ are part of the galenobismutite, lillianite and pavonite homologous series respectively (Mrotzek and Kanatzidis 2003b; Makovicky 2006; Zhao et al. 2019a). Table 4.2 summaries the thermoelectric properties of $PbBi_2S_4$, $Pb_3Bi_2S_6$, $InPbBi_3S_7$ and $CdPb_2Bi_4S_9$ (Ohta et al. 2014; Zhao et al. 2019a; Cai et al. 2021). The negative sign of the Seebeck coefficient confirms n-type (electron) carrier transport in all samples. Moreover, both electrical resistivity and the Seebeck coefficient increase monotonically with temperature for all samples, which is consistent with a degenerate semiconducting behavior. $PbBi_2S_4$ was discovered to have a high power factor of ~370 µW m^{-1} K^{-2} at 800 K. The complex structure of these homologous compounds yields low lattice thermal conductivity, resulting in high zT (see Table 4.2). The lattice thermal conductivity and zT of PbBi$_2$S$_4$ were ~0.4 W·m^{-1}·K^{-1} and ~0.46 at 800 K respectively (Cai et al. 2021). The crystal structural evolution would optimize the electrical and thermal transport properties in homologous series of compounds (Mrotzek and Kanatzidis 2003a; Kanatzidis 2005), increasing the zT value.

Materials	T (K)	ρ (μΩ m)	S (μW K^{-1})	S^2/ρ (μW m^{-1} K^{-2})	κ (W m^{-1} K^{-1})	κ_l (W m^{-1} K^{-1})	zT	References
PbBi$_2$S$_4$	710	270	−270	260	0.57	0.5	0.33	Ohta et al. (2014)
PbBi$_2$S$_4$	800	54	−142	370	0.65	0.4	0.46	Cai et al. (2021)
Pb$_3$Bi$_2$S$_6$	720	150	−200	260	0.67	0.6	0.26	Ohta et al. (2014)
Pb$_3$Bi$_2$S$_6$	800	40	−100	250	1.0	0.6	0.20	Cai et al. (2021)
CdPb$_2$Bi$_4$S$_9$	775	270	−250	230	0.53	0.5	0.53	Zhao et al. (2019a)

Table 4.2. *Electrical resistivity, ρ, Seebeck coefficient, S, power factor, S^2/ρ, total thermal conductivity, κ, lattice thermal conductivity, κ_l and thermoelectric figure of merit, zT, of PbBi$_2$S$_4$, Pb$_3$Bi$_2$S$_6$ and CdPb$_2$Bi$_4$S$_9$*

4.2.4. Cu and Ag sulfide-based superionic conductors

Ag$_2$S is a superionic conductor with a bandgap around 1.1 eV (at room temperature) (Kashida et al. 2003); however, the mixed conduction of electrons and Ag ions is mainly prevalent above 451 K (in high-temperature cubic phase) (Miyatami 1968). The room temperature modification of Ag$_2$S (α-phase) is monoclinic with space group $P2_1/c$ and is stable up to 451 K, above which it undergoes a transition to a body-centered cubic structure, which is stable between 449 K and 895 K (Shi et al. 2018; Zhou et al. 2020). A face-centered cubic structure is formed at higher temperatures. The crystal structure in silver chalcogenides can be tuned to obtain the desired mechanical and thermoelectric properties, allowing for the formation of an optimized thermoelectric material (Liang et al. 2019; Jood et al. 2020). For example, depending on the S/Se atomic ratio, Ag$_2$(S, Se) compound crystallizes into an orthorhombic or monoclinic crystal structure with different mechanical and electrical properties (Liang et al. 2020).

Pristine Ag_2S is limited by its low electron concentration (order of 10^{14} cm^{-3}) at room temperature, which has recently been significantly improved (Liang et al. 2019, 2020; Wang et al. 2021a). Se and Te alloying minimize the defect formation energy of Ag interstitials, which significantly improves the electrical conductivity and zT in the 300–450 K range, making them promising for room temperature applications (Liang et al. 2019). For example, zT of ~0.44 for $Ag_2S_{0.5}Se_{0.45}Te_{0.05}$ and ~0.63 for $Ag_2S_{0.8}Te_{0.2}$ at 300 K and 450 K, respectively, were achieved compared with $zT < 0.001$ for pristine Ag_2S at 300 K. Recently, a new ductile semiconductor, $Ag_{20}S_7Te_3$, was found to have a zT of ~0.80 at 600 K while still maintaining good flexibility and plastic deformability as monoclinic α-Ag_2S (Yang et al. 2021). In contrast to other inorganic semiconductors, α-Ag_2S has metal-like ductility with significant plastic deformation stresses (~4.2% elongation for tension, above 50% for compression and above 20% for bending) due to its distinctive structural and chemical bonding properties (Shi et al. 2018). Silver chalcogenides-based compounds are extremely appealing for flexible thermoelectric device development due to their unique mechanical properties and strong thermoelectric performance, as mentioned in a later section.

$Cu_{2-x}S$ is a superionic mixed ionic–electronic conductor investigated for decades, particularly for its complex structure and solar cell applications due to its relatively small bandgap around 1.2 eV. Copper sulfide has complicated crystal structures at a low-temperature and two phase transitions at 370 K and 700 K (Chakrabarti and Laughlin 1983). Copper sulfide crystal structure is a monoclinic low chalcocite γ-phase (L-Chalcocite) at 370 K. It becomes hexagonal high chalcocite β-phase (H-Chalcocite) in the temperature range of 370–700 K, which has been reported as a solid–liquid hybrid phase with copper in the liquid-like substructure. Furthermore, when the temperature is higher than 700 K, it transforms into the face-centered cubic (fcc) α-phase, which is a superionic phase that has mobile copper ions in the structure. These liquid-like copper ions in a sulfide sublattice play a significant role in achieving an extremely low lattice thermal conductivity, resulting in high thermoelectric performance.

$Cu_{2-x}S$ has recently attracted attention as thermoelectric materials with the "phonon-liquid electron-crystal" (PLEC) concept (Liu et al. 2012; Dennler et al. 2014; Ge et al. 2016; Zhao et al. 2019b). Cu_2S is a major research subject in the field of thermoelectrics due to its intriguing thermal and electrical transport properties, such as reduced specific heat and extremely low lattice thermal conductivity (below 0.35 W m^{-1} K^{-1}). Although ideal Cu_2S is an intrinsic semiconductor, the sample with a nominal Cu_2S composition shows p-type behavior due to the Cu deficiency. The hole carrier concentration can be tuned by changing the Cu content, and the thermoelectric figure of merit, zT, of $Cu_{2-x}S$ can reach above 1.5. For example, zT was improved from ~0.65 for Cu_2S to ~1.7 for $Cu_{1.97}S$ at 1,000 K (He et al. 2014).

4.2.5. Tetrahedrites and colusites

Tetrahedrites ($Cu_{12}Sb_4S_{13}$) and colusites ($Cu_{26}A_2E_6S_{32}$; A: V, Nb, Ta, Cr, Mo, W; E: Sn, Ge) have received significant attention as emerging p-type thermoelectric materials. Both systems are mainly composed of the earth-abundant and low-toxicity elements, Cu and S. The tetrahedrites possess a cubic structure (space group: $I\bar{4}3m$), consisting of a CuS_4 tetrahedron, CuS_3 trigonal planar unit and SbS_3 trigonal pyramid (Suekuni and Takabatake 2016). The colusites is composed of numerous atoms (66) in the cubic unit cell ($P\bar{4}3n$) (Suekuni and Takabatake 2016), consisting of AS_4 tetrahedron, CuS_4 tetrahedra and ES_4 tetrahedra (Suekuni and Takabatake 2016).

Tetrahedrites and colusites have a high power factor because their valence bands are mainly composed of hybridized Cu 3D and S 3P orbitals (Suekuni et al. 2014; Lu et al. 2015; Suekuni and Takabatake 2016). In tetrahedrites, three-fold coordinated Cu atoms vibrate anharmonically perpendicular to the triangular plane of S atoms (Suekuni et al. 2013, 2018; Lu et al. 2013). The lattice thermal conductivity was suppressed by this out-of-plane vibration. In colusites, the low lattice thermal conductivity is due to the structural complexity and low-frequency vibration of Cu atoms (Suekuni et al. 2014; Bourgès et al. 2018). The lattice thermal conductivity is further reduced by introducing various types of defects into the crystal structure, such as interstitial defects, antisite defects between cations and split Cu sites (Kikuchi et al. 2016; Bourgès et al. 2018; Suekuni et al. 2019; Bouyrie et al. 2020; Shimizu et al. 2021). These defects are formed through sulfur losses when the samples are heated above 973 K.

Table 4.3 summarizes several significant breakthroughs in the thermoelectric properties of tetrahedrites and colusites. For tetrahedrites, a high power factor and low lattice thermal conductivity were observed at room temperature for $Cu_{10}Tr_2Sb_4S_{13}$ (Tr: Mn, Fe, Co, Ni, Cu and Zn) in 2012 (Suekuni et al. 2012). After a year, high zT values were obtained at a high temperature; for example, zT of ~0.7 at 665 K and ~0.95 at 700 K have been obtained for $Cu_{10.5}Ni_{1.5}Sb_4S_{13}$ and $Cu_{11.5}Zn_{0.5}Sb_4S_{13}$ respectively (Lu et al.

2013; Suekuni et al. 2013). The zT value was further enhanced to over 1.0 by tuning the chemical composition (Lu et al. 2015; Yan et al. 2018) and inserting a porous network in sintered samples (Hu et al. 2021). For colusites, $Cu_{26}V_2Ge_6S_{32}$ was reported in 2014 to have a zT of ~0.73 at 660 K due to a high power factor and low lattice thermal conductivity (Suekuni et al. 2014). The zT value was recently improved to over 1.0 by tuning the carrier concentration and introducing various types of defects into the crystal structure (Bouyrie et al. 2017).

Material	T (K)	ρ (μΩ m)	S (μW K^{-1})	S^2/ρ (μW m^{-1} K^{-2})	κ (W m^{-1} K^{-1})	zT	References
Tetrahedrite							
$Cu_{10.5}Ni_{1.5}Sb_4S_{13}$	665	40	190	900	0.8	0.7	Suekuni et al. (2013)
$Cu_{11.5}Zn_{0.5}Sb_4S_{13}$	730	20	160	1,380	1.1	0.95	Lu et al. (2013)
$Cu_{10.5}Ni_{1.0}Zn_{0.5}Sb_4S_{13}$	720	55	220	850	0.6	1.03	Lu et al. (2015)
$Cu_{13.5}Sb_4S_{12}Se$	720	25	170	1,200	0.8	1.1	Yan et al. (2018)
Porous $Cu_{12}Sb_4Sb_{13}$	720	20	160	1,300	0.8	1.15	Hu et al. (2021)
Colusite							
$Cu_{26}V_2Ge_6S_{32}$	660	70	210	620	0.6	0.73	Suekuni et al. (2014)
$Cu_{26}V_2Sn_6S_{32}$	680	35	160	750	0.5	0.93	Bourgès et al. (2018)
$Cu_{26}Ta_2Sn_{5.5}S_{32}$	670	55	210	800	0.5	1.0	Bouyrie et al. (2017)
$Cu_{26}Cr_2Ge_6S_{32}$	700	12	150	1,940	1.6	0.86	Kumar et al. (2019)
$Cu_{29}V_2Ge_5SbS_{32}$	670	45	200	950	0.8	0.8	Shimizu et al. (2021)

Table 4.3. *Major milestones achieved for thermoelectric properties of Cu–S-based tetrahedrites and colusites. Electrical resistivity, ρ, Seebeck coefficient, S, power factor, S^2/ρ, total thermal conductivity, κ and thermoelectric figure of merit zT are listed*

4.2.6. Chevrel-phase sulfides

Chevrel-phase chalcogenides have attracted significant attention for many years due to their remarkable superconducting properties, including a large critical magnetic field and high critical superconducting transition temperature (Perrin et al. 2019). The interest in the thermoelectric properties of Chevrel-phase chalcogenides has recently increased due to the high zT at high temperatures (Roche et al. 1998; Caillat et al. 1999; Tsubota et al. 1999; Zhou et al. 2011; Gougeon et al. 2012). The general formula of Chevrel-phase chalcogenides is $M_xMo_6Ch_8$ (M: Metal; Ch: S, Se and Te). The crystal structure consists of stacked Mo_6Ch_8 building blocks. Metal, M, ions fill the cavities between the Mo_6Ch_8 building blocks.

Materials	T (K)	ρ (μΩ m)	S (μW K^{-1})	S^2/ρ (μW m^{-1} K^{-2})	κ (W m^{-1} K^{-1})	zT	References
$Cu_{2.5}Mo_6S_8$	950	15	75	440	2.5	0.2	Ohta et al. (2009c)
$Cu_{4.0}Mo_6S_8$	950	20	130	810	2.0	0.4	Ohta et al. (2009c)
$Cr_{1.3}Mo_6S_8$	970	10	70	570	3.0	0.17	Ohta et al. (2010)
$Mn_{1.3}Mo_6S_8$	970	55	95	330	3.0	0.11	Ohta et al. (2010)
$Ag_3Tl_2Mo_{15}S_{19}$	800	17	70	260	1.5	0.15	Gougeon et al. (2021)

Table 4.4. *Recently reported electrical resistivity, ρ, Seebeck coefficient, S, power factor, S^2/ρ, total thermal conductivity, κ and thermoelectric figure of merit, zT, of Chevrel-phase sulfides*

Table 4.4 summarizes the recently reported thermoelectric properties of Chevrel-phase sulfides (Ohta et al. 2009c, 2010; Gougeon et al. 2021). The optimization in the guest M content results in the tuning of carrier concentration, boosting the high power factor. For $Cu_xMo_6S_8$, the power factor was boosted from ~440 μW K^{-2} m^{-1} for $x = 2.5$ to ~810 μW K^{-2} m^{-1} for $x = 4.0$ at 950 K. The low lattice thermal conductivity is most likely due to the rattling of guest M ions. The lattice thermal conductivity of $Cu_xMo_6S_8$ at 950 K is ~2.0 W m^{-1} K^{-1}. The high power factor and low lattice thermal conductivity result in high zT in the Chevrel-phase sulfides. The zT obtained for $Cu_{4.0}Mo_6S_8$ at 950 K is ~0.4. The building blocks evolve from Mo_6S_8 to Mo_9S_{11} and $Mo_{12}S_{14}$, and can be represented by the general formula $Mo_{3n}S_{3n+2}$ ($n \geq 2$) (Perrin et al. 2019). The giant building block is expected to improve zT by reducing the low lattice thermal conductivity. The lattice thermal conductivity of $Ag_3Tl_2Mo_{15}S_{19}$ at 950 K is ~1.5 W m^{-1} K^{-1}. (Gougeon et al. 2021). However, only a few research works on the thermoelectric properties of $n > 2$ systems have been reported. The $n \geq 2$ systems would be good areas to seek for high zT materials.

4.2.7. Chalcopyrite

Chalcopyrite is a compound with the ABX_2 (A = Ag, Cu; B = Al, Ga, In, Fe, Tl; X = S, Se, Te) composition (Austin et al. 1956; Jaffe and Zunger 1983; Li et al. 2014). As sulfide materials, n-type $CuFeS_2$ (Li et al. 2013) and p-type $CuAlS_2$, (Liu et al. 2007a, 2007b) have been studied as potential thermoelectric materials composed of cost-effective and earth-abundant elements. A natural mineral composed of chalcopyrite $CuFeS_2$ that exists at a deep-sea hydrothermal vent can demonstrate thermoelectricity as proof of the environmentally friendly characteristics of chalcopyrite phases (Ang et al. 2015).

$CuFeS_2$ can be synthesized chemically or by combining mechanical alloying and high-temperature densification (Li et al. 2013, 2014). The system is expected to be suitable for low- to mid-temperature thermoelectric applications, with a bandgap of ~0.53 eV (Austin et al. 1956) or 0.3 eV (Hamajima et al. 1981) determined by optical absorption measurement and calculation respectively. In terms of $CuFeS_2$ stability, thermogravimetry–differential thermal analysis showed the decomposition of the chalcopyrite phase into an isometric (Cu, Fe)S phase and pyrite FeS_2 around 800 K (Tsujii et al. 2014). The power factor of n-type $CuFeS_2$ was enhanced to ~1 mW m^{-1} K^{-2} via Zn doping on Cu site or optimizing Fe content in the range of 400–600 K (Tsujii et al. 2014). $CuFeS_{1.80}$ and Mn/Co-doped compositions achieved a maximum zT value of ~ 0.2 at 600 K (Li et al. 2013; Lefèvre et al. 2016). In comparison to other systems with ionic conduction issues, $CuFeS_2$ has a substantially lower ionic conduction with Cu migration (Wang et al. 2021b). Furthermore, $CuFeS_2$ bulk materials have intriguing optical and magnetic properties. Magnetism has been indicated to enhance the thermoelectric properties (Tsujii and Mori 2013).

Because p-type $CuAlS_2$ has the widest bandgap (E_g) among ABX_2 chalcopyrite compounds, it has been explored as a promising light-emitting material or transparent semiconductor for a long time (Liu et al. 2007a, 2007b). The bandgap E_g (~3.0 eV) of $CuAlS_2$ estimated from the optical absorption spectrum decreases as the dopant concentration of Zn increases (Liu et al. 2007a). Although n-type $CuFeS_2$ still has a substantially low zT value, several efforts have been made to improve its thermoelectric performance, such as making a composite with carbon nanotube (Shojaei et al. 2022).

4.3. Modules

Sulfide-based thermoelectric power generation modules are now being fabricated. The most critical factor in fabricating modules is to achieve stable operation under temperature differences. The temperature gradient in the module can cause a chemical reaction and/or atomic diffusion in the thermoelectric legs and/or at the interface between the electrodes and the legs under the temperature difference, resulting in module degradation and conversion efficiency reduction (Zebarjadi et al. 2012; Zhang et al. 2016; He et al. 2018; Tan et al. 2019). Efforts have been made to improve the module's stability.

4.3.1. *Colusites*

This section describes the advancements in the fabrication of power generation devices based on colusites. The $Cu_{26}Nb_2Ge_6S_{32}$-based elements with metal contact layers (Ti, Pt, Ni or Au) were fabricated via hot pressing (Chetty et al. 2019). Microcracks formed at the interface between $Cu_{26}Nb_2Ge_6S_{32}$ and Ti or Pt due to a mismatch in their coefficients of thermal expansion, resulting in increased interface resistance and lower conversion efficiency. $Cu_{26}Nb_2Ge_6S_{32}$ (Chetty et al. 2019), Ti (Touloukian et al. 1975) and Pt (Touloukian et al. 1975) have thermal expansions of ~16.9 × 10^{-6} K^{-1}, ~10.4×10^{-6} K^{-1} and ~9.8 × 10^{-6} K^{-1}, respectively, at 573 K. Because the coefficients of thermal expansion between $Cu_{26}Nb_2Ge_6S_{32}$ and the Ni or Au contact layers were close (Ni: ~15.9 × 10^{-6} K^{-1} at 573 K (Touloukian et al. 1975), Au: ~16.9 × 10^{-6} K^{-1} at 573 K (Touloukian et al. 1975)), no cracks occurred at the interface. However, in the case of the Ni contact layer, secondary phases were formed around the interface, resulting in high interface resistance and reduced conversion efficiency. In the case of the Au contact layer, no secondary phases were found around the interface, allowing a reduced specific contact resistance of 4×10^{-10} to 5×10^{-10} Ω m.

Table 4.5 summarizes the power generation characteristics of the $Cu_{26}Nb_2Ge_6S_{32}$-based elements. For temperature differences of 270 K, the maximum thermoelectric conversion efficiency η_{max} was estimated to be ~3.3% in the $Cu_{26}Nb_2Ge_6S_{32}$-based thermoelectric element with the Au-based contact layer. The prototype $Cu_{26}Cr_2Ge_6S_{32}$-based thermoelectric power generation module was developed alongside n-type $Pb_{0.98}Ga_{0.02}Te$-3% GeTe (Chetty et al. 2022). When the hot- and cold-side temperature were maintained at 673 and 283 K, respectively, η_{max} of ~5.5% was obtained for the $Cu_{26}Cr_2Ge_6S_{32}/Pb_{0.98}Ga_{0.02}Te$-3% GeTe module.

T_h (K)	R_{in} (mΩ)	P_{max} (mW)	η_{max} (%)
370	11.8	1.15	0.8
470	13.4	6.84	2.0
570	18.0	17.8	3.3

Table 4.5. *Internal resistance (R_{in}), maximum output power (P_{max}) and maximum conversion efficiency (η_{max}) of the $Cu_{26}Nb_2Ge_6S_{32}$ single element of ~ 3 mm × ~ 3 mm and a height of 5 mm. The cold-side temperature (T_c) was maintained at 297 K, whereas the hot-side temperature (T_h) was varied from 370 to 570 K (Chetty et al. 2019)*

4.3.2. Cu and Ag sulfide-based superionic conductors

The effects of mobile Cu ions must be considered when fabricating thermoelectric modules with Cu_2S (or more generally with the superionic mixed ionic–electronic conductors) (Qiu et al. 2018, 2019). The diffusivity of mobile ions (Cu^+ ions in Cu_2S) is low due to its low activation energy, E_a (e.g. 0.19 eV in Cu_2S) (Balapanov et al. 2004). Energetically, the neighboring atomic site is close, and ions can jump frequently to the neighboring sites. The charge carrier flows in a specific direction in response to an external driving force, such as an electric field or temperature gradient. These phenomena can cause the deposition of Cu on the surface and the degradation of thermoelectric legs.

Cu metal deposited on the sample surface can damage the electrode contact by causing cracks or incoherent interfaces. These changes increase the electrical and thermal resistance at the interface and degrade the performance of the thermoelectric module (Dennler et al. 2014; Powell 2019). In addition, Cu movement towards the cathode may result in the evaporation of chalcogen at the anode. In a similar Cu_2Se-based material, NASA Jet Propulsion Laboratory stopped the development of radioisotope thermal generators for space applications in 1979, after more than 10 years of research due to these instabilities. Therefore, this degradation behavior must be addressed before employing in any applications (Brown et al. 2013; Dennler et al. 2014). Several trials have been proposed to address these problems, including the use of ion blocking; however, electronically conductive interfaces realize a high total voltage with large current densities (Qiu et al. 2018, 2019). Based on the thermodynamic explanation, critical voltage rather than current density is the deciding factor for Cu deposition. Therefore, a thermoelectric leg segmented by ion blocking layers can prevent Cu migration and improve stability. This study shows that if the leg's chemical potential or segmented portion is selected to be lower than the corresponding critical voltage (chemical potential), Cu-based compounds (and other superionic mixed ionic–electronic conductors) can effectively provide larger current densities and have high-performance with stability.

Ag migration in Ag-based chalcogenides has not been studied in terms of module development, unlike Cu_2(S, Se). However, the few reports on their power generating characteristics are promising. It has been recently shown that Ag_2S-based compounds can be directly cut into flexible thin foils from bulk samples, allowing for the development of flexible thermoelectric generators for wearables without the support of substrates (Liang et al. 2019). This is due to their unique flexibility and metallic plastic deformability (Shi et al. 2018). A six-couple flexible device with a lateral Π-shaped configuration using $Ag_2S_{0.5}Se_{0.5}$ foils (n-type legs) was fabricated, in which Pt–Rh wires were used to connect the top and bottom ends of the thermoelectric legs in series. This module generated a power output of ~10 μW at temperature difference of 20 K (cold-side temperature of 293 K) (Liang et al. 2019). In another study, a cylindrical hetero-shaped thermoelectric module using strips of $Ag_{20}S_7Te_3$, which is also a superionic conductor with excellent shape-conformability, was fabricated (Yang et al. 2021). A power output of ~17.1 μW was achieved under temperature difference of 70 K.

4.4. Summary and prospects

This chapter has addressed the recent progress in selected thermoelectric sulfides and their modules. The high thermoelectric figure of merit, zT, in sulfide materials was enhanced using novel strategies, including carrier concentration tuning, electronic band engineering, crystal structure evolution and phonon engineering. Advanced sulfide research has progressed from materials development to module fabrication. In addition, power generation in sulfide-based modules has been demonstrated. For further progress and social implementation of sulfide thermoelectrics, more effort and progress should be devoted to improving the module performance, stability, mechanical strength and system assembly for the social implementation of sulfide thermoelectrics.

4.5. References

Ang, R., Khan, A.U., Tsujii, N., Takai, K., Nakamura, R., Mori, T. (2015). Thermoelectricity generation and electron-magnon scattering in a natural chalcopyrite mineral from a deep-sea hydrothermal vent. *Angew. Chemie – Int. Ed.*, 54(44), 12909–12913. https://doi.org/10.1002/anie.201505517.

Austin, I.G., Goodman, C.H.L., Pengelly, A.E. (1956). New semiconductors with the chalcopyrite structure. *J. Electrochem. Soc.*, 103(11), 609–610. https://doi.org/10.1149/1.2430171.

Balapanov, M.Kh., Gafurov, I.G., Mukhamed'yanov, U.Kh., Yakshibaev, R.A., Ishembetov, R.K. (2004). Ionic conductivity and chemical diffusion in superionic $Li_xCu_{2-x}S$ ($0 \leq x \leq 0.25$). *Phys. Status Solidi Basic Res.*, 241(1), 114–119. https://doi.org/10.1002/pssb.200301911.

Barbier, T., Lebedev, O.I., Roddatis, V., Bréard, Y., Maignan, A., Guilmeau, E. (2015). Silver intercalation in SPS dense TiS_2: Staging and thermoelectric properties. *Dalt. Trans.*, 44(17), 7887–7895. https://doi.org/10.1039/C5DT00551E.

Beaumale, M., Barbier, T., Bréard, Y., Guelou, G., Powell, A.V., Vaqueiro, P., Guilmeau, E. (2014). Electron doping and phonon scattering in $Ti_{1+x}S_2$ thermoelectric compounds. *Acta Mater.*, 78, 86–92. https://doi.org/https://doi.org/10.1016/j.actamat.2014.06.032.

Biswas, K., Zhao, L.D., Kanatzidis, M.G. (2012). Tellurium-free thermoelectric: The anisotropic *n*-Type semiconductor Bi_2S_3. *Adv. Energy Mater.*, 2(6), 634–638. https://doi.org/10.1002/aenm.201100775.

Bourgès, C., Bouyrie, Y., Supka, A.R., Al Orabi, R.A., Lemoine, P., Lebedev, O.I., Ohta, M., Suekuni, K., Nassif, V., Hardy, V. et al. (2018). High-performance thermoelectric bulk colusite by process controlled structural disordering. *J. Am. Chem. Soc.*, 140(6), 2186–2195. https://doi.org/10.1021/jacs.7b11224.

Bouyrie, Y., Ohta, M., Suekuni, K., Kikuchi, Y., Jood, P., Yamamoto, A., Takabatake, T. (2017). Enhancement in the thermoelectric performance of colusites $Cu_{26}A_2E_6S_{32}$ (A = Nb, Ta; E = Sn, Ge) using E-Site non-stoichiometry. *J. Mater. Chem. C*, 5(17), 4174–4184. https://doi.org/10.1039/C7TC00762K.

Bouyrie, Y., Chetty, R., Suekuni, K., Saitou, N., Jood, P., Yoshizawa, N., Takabatake, T., Ohta, M. (2020). Enhancement of the thermoelectric power factor by tuning the carrier concentration in Cu-Rich and Ge-Poor colusites $Cu_{26+x}Nb_2Ge_{6-x}S_{32}$. *J. Mater. Chem. C*, 8(19), 6442–6449. https://doi.org/10.1039/D0TC00508H.

Brown, D.R., Day, T., Caillat, T., Snyder, G.J. (2013). Chemical stability of $(Ag,Cu)_2Se$: A historical overview. *J. Electron. Mater*, 42(7), 2014–2019. https://doi.org/10.1007/s11664-013-2506-2.

Cai, F.G., Dong, R., Sun, W., Lei, X.B., Yu, B., Chen, J., Yuan, L., Wang, C., Zhang, Q.Y. (2021). $Pb_mBi_2S_{3+m}$ homologous series with low thermal conductivity prepared by the solution-based method as promising thermoelectric materials. *Chem. Mater.*, 33(15), 6003–6011. https://doi.org/10.1021/acs.chemmater.1c01387.

Caillat, T., Fleurial, J.-P., Snyder, G.J. (1999). Potential of Chevrel phases for thermoelectric applications. *Solid State Sci.*, 1(7–8), 535–544. https://doi.org/10.1016/S1293-2558(00)80105-3.

Chakrabarti, D.J. and Laughlin, D.E. (1983). The Cu-S (Copper-Sulfur) System. *Bull. Alloy Phase Diagrams*, 4(3), 254–271. https://doi.org/10.1007/BF02868665.

Chetty, R., Kikuchi, Y., Bouyrie, Y., Jood, P., Yamamoto, A., Suekuni, K., Ohta, M. (2019). Power generation from the $Cu_{26}Nb_2Ge_6S_{32}$-based single thermoelectric element with au diffusion barrier. *J. Mater. Chem. C*, 7(17), 5184–5192. https://doi.org/10.1039/C9TC00868C.

Chetty, R., Jood, P., Murata, M., Suekuni, K., Ohta, M. (2022). Prototype thermoelectric module based on p-Type colusite together with n-Type nanostructured PbTe for power generation. *Appl. Phys. Lett.*, 120(1), 013501, 1–7. https://doi.org/10.1063/5.0077154.

Cutler, M., Leavy, J.F., Fitzpatrick, R.L. (1964). Electronic transport in semimetallic cerium sulfide. *Phys. Rev.*, 133(4A), A1143–A1152. https://doi.org/10.1103/PhysRev.133.A1143.

Dennler, G., Chmielowski, R., Jacob, S., Capet, F., Roussel, P., Zastrow, S., Nielsch, K., Opahle, I., Madsen, G.K.H. (2014). Are binary copper sulfides/selenides really new and promising thermoelectric materials? *Adv. Energy Mater.*, 4(9), 1301581, 1–12. https://doi.org/10.1002/aenm.201301581.

Flahaut, J. (1979). Sulfides, selenides and tellurides. In *Handbook on the Physics and Chemistry of Rare Earths*, Gschneidner, K.A. and Eyring, L. (eds). North-Holland Publishing Company, Amsterdam. https://doi.org/10.1016/S0168-1273(79)04004-6.

Gadzhiev, G.G., Ismailov, S.M., Khamidov, M.M., Abdullaev, Kh.Kh., Sokolov, V.V. (2000). Thermophysical properties of sulfides of lanthanum, praseodymium, gadolinium, and dysprosium. *High Temp.*, 38(6), 875–879. https://doi.org/10.1023/A:1004185105712.

Ge, Z.H., Liu, X.Y., Feng, D., Lin, J.Y., He, J.Q. (2016). High-performance thermoelectricity in nanostructured earth-abundant copper sulfides bulk materials. *Adv. Energy Mater.*, 6(16), 1600607, 1–7. https://doi.org/10.1002/aenm.201600607.

Girard, S.N., He, J.Q., Zhou, X.Y., Shoemaker, D., Jaworski, C.M., Uher, C., Dravid, V.P., Heremans, J.P., Kanatzidis, M.G. (2011). High performance Na-doped PbTe–PbS thermoelectric materials: Electronic density of states modification and shape-controlled nanostructures. *J. Am. Chem. Soc.*, 133(41), 16588–16597. https://doi.org/10.1021/ja206380h.

Gougeon, P., Gall, P., Al Orabi, R.A., Fontaine, B., Gautier, R., Potel, M., Zhou, T., Lenoir, B., Colin, M., Candolfi, C. et al. (2012). Synthesis, crystal and electronic structures, and thermoelectric properties of the novel cluster compound $Ag_3In_2Mo_{15}Se_{19}$. *Chem. Mater.*, 24(15), 2899–2908. https://doi.org/10.1021/cm3009557.

Gougeon, P., Gall, P., Misra, S., Dauscher, A., Candolfi, C., Lenoir, B. (2021). Synthesis, crystal structure and transport properties of the cluster compounds $Tl_2Mo_{15}S_{19}$ and $Ag_3Tl_2Mo_{15}S_{19}$. *Mater. Res. Bull.*, 136, 111152, 1–9. https://doi.org/10.1016/j.materresbull.2020.111152.

Gschneidner, K.A. (1992). The paper "Pr-S (Praseodymium-Sulfur)". *J. Phase Equilibria*, 13(6), 586–587. https://doi.org/10.1007/BF02667205.

Gschneidner, K.A. (1998). Preparation and processing of rare earth chalcogenides. *J. Mater. Eng. Perform.*, 7(5), 656–660. https://doi.org/10.1361/105994998770347521.

Gschneidner, K.A., Nakahara, J.F., Beaudry, B.J., Takeshita, T., Ames Laboratory (1987). Lanthanide refractory semiconductors based on the Th_3P_4 structure. *MRS Proc.*, 97, 359–370. https://doi.org/10.1557/PROC-97-359.

Gu, Y., Song, K.K., Hu, X.H., Chen, C.C., Pan, L., Lu, C.H., Shen, X.D., Koumoto, K., Wang, Y.F. (2020). Realization of an ultrahigh power factor and enhanced thermoelectric performance in TiS_2 via microstructural texture engineering. *ACS Appl. Mater. Interfaces*, 12(37), 41687–41695. https://doi.org/10.1021/acsami.0c09592.

Guélou, G., Vaqueiro, P., Prado-Gonjal, J., Barbier, T., Hébert, S., Guilmeau, E., Kockelmann, W., Powell, A.V. (2016). The impact of charge transfer and structural disorder on the thermoelectric properties of cobalt intercalated TiS_2. *J. Mater. Chem. C*, 4(9), 1871–1880. https://doi.org/10.1039/C5TC04217H.

Guélou, G., Lemoine, P., Raveau, B., Guilmeau, E. (2021). Recent developments in high-performance thermoelectric sulphides: An overview of the promising synthetic colusites. *J. Mater. Chem. C*, 9(3), 773–795. https://doi.org/10.1039/d0tc05086e.

Guilmeau, E., Bréard, Y., Maignan, A. (2011). Transport and thermoelectric properties in copper intercalated TiS_2 chalcogenide. *Appl. Phys. Lett.*, 99(5), 052107, 1–3. https://doi.org/10.1063/1.3621834.

Guilmeau, E., Barbier, T., Maignan, A., Chateigner, D. (2017). Thermoelectric anisotropy and texture of intercalated TiS_2. *Appl. Phys. Lett.*, 111(13), 133903, 1–4. https://doi.org/10.1063/1.4998952.

Guittard, M. and Flahaut, J. (1991). Preparation of rare earth sulfides and selenides. In *Synthesis of Lanthanide and Actinide Compounds*, Meyer, G. and Morss, L.R. (eds). Kluwer Academic Publishers, Amsterdam. https://doi.org/10.1007/978-94-011-3758-4_14.

Guo, J., Lou, Q., Qiu, Y., Wang, Z.Y., Ge, Z.H., Feng, J., He, J.Q. (2020a). Remarkably enhanced thermoelectric properties of Bi_2S_3 nanocomposites via modulation doping and grain boundary engineering. *Appl. Surf. Sci.*, 520, 146341, 1–8. https://doi.org/10.1016/j.apsusc.2020.146341.

Guo, J., Zhang, Y.X., Wang, Z.Y., Zheng, F.S., Ge, Z.H., Fu, J., Feng, J. (2020b). High thermoelectric properties realized in earth-abundant Bi_2S_3 bulk via carrier modulation and multi-nano-precipitates synergy. *Nano Energy*, 78, 105227, 1–8. https://doi.org/10.1016/j.nanoen.2020.105227.

Hamajima, T., Kambara, T., Gondaira, K.I., Oguchi, T. (1981). Self-consistent electronic structures of magnetic semiconductors by a discrete variational $X\alpha$ calculation III: Chalcopyrite $CuFeS_2$. *Phys. Rev. B*, 24(6), 3349–3353. https://doi.org/10.1103/PhysRevB.24.3349.

Han, S.H. and Cook, B.A. (1994). An experimental search for high *ZT* semiconductors: A survey of the preparation and properties of several alloy systems. *AIP Conference Proceedings*, AIP, 316, 66–70. https://doi.org/10.1063/1.46837.

He, Y., Day, T., Zhang, T.S., Liu, H.L., Shi, X., Chen, L.D., Snyder, G.J. (2014). High thermoelectric performance in non-toxic earth-abundant copper sulfide. *Adv. Mater.*, 26(23), 3974–3978. https://doi.org/10.1002/adma.201400515.

He, R., Schierning, G., Nielsch, K. (2018). Thermoelectric devices: A review of devices, architectures, and contact optimization. *Adv. Mater. Technol.*, 3(4), 1700256, 1–17. https://doi.org/10.1002/admt.201700256.

Hébert, S., Berthebaud, D., Daou, R., Bréard, Y., Pelloquin, D., Guilmeau, E., Gascoin, F., Lebedev, O., Maignan, A. (2016). Searching for new thermoelectric materials: Some examples among oxides, sulfides and selenides. *J. Phys. Condens. Matter*, 28(1), 013001, 1–23. https://doi.org/10.1088/0953-8984/28/1/013001.

Henderson, J.R., Muramoto, M., Loh, E., Gruber, J.B. (1967). Electronic structure of rare-earth sesquisulfide crystals. *J. Chem. Phys.*, 47(9), 3347–3356. https://doi.org/10.1063/1.1712397.

Hirai, S., Shimakage, K., Saitou, Y., Nishimura, T., Uemura, Y., Mitomo, M., Brewer, L. (1998). Synthesis and sintering of cerium(III) sulfide powders. *J. Am. Ceram. Soc.*, 81(1), 145–151. https://doi.org/10.1111/j.1151-2916.1998.tb02306.x.

Hirai, S., Suzuki, K., Shimakage, K., Nishimura, S., Uemura, Y., Mitomo, M. (2003). Preparations of γ-Pr_2S_3 and γ-Nd_2S_3 powders by sulfurization of Pr_6O_{11} and Nd_2O_3 powders using CS_2 gas, and their sintering. *J. Jpn. Inst. Met.*, 67(1), 15–21. https://doi.org/10.2320/jinstmet1952.67.1_15.

Hu, H.H., Zhuang, H.L., Jiang, Y.L., Shi, J.L., Li, J.W., Cai, B.W., Han, Z.R., Pei, J., Su, B., Ge, Z.H. et al. (2021). Thermoelectric $Cu_{12}Sb_4S_{13}$-based synthetic minerals with a sublimation-derived porous network. *Adv. Mater*, 2103633, 1–10. https://doi.org/10.1002/adma.202103633.

Imai, H., Shimakawa, Y., Kubo, Y. (2001). Large thermoelectric power factor in TiS_2 crystal with nearly stoichiometric composition. *Phys. Rev. B*, 64(24), 241104, 1–4. https://doi.org/10.1103/PhysRevB.64.241104.

Jaffe, J.E. and Zunger, A. (1983). Electronic structure of the ternary chalcopyrite semiconductors $CuAlS_2$, $CuGaS_2$, $CuInS_2$, $CuAlSe_2$, $CuGaSe_2$, and $CuInSe_2$. *Phys. Rev. B*, 28(10), 5822–5847. https://doi.org/10.1103/PhysRevB.28.5822.

Ji, W.T., Shi, X.L., Liu, W.D., Yuan, H.L., Zheng, K., Wan, B.A., Shen, W.X., Zhang, Z.F., Fang, C., Wang, Q.Q. et al. (2021). Boosting the thermoelectric performance of n-type Bi_2S_3 by hierarchical structure manipulation and carrier density optimization. *Nano Energy*, 87, 106171, 1–11. https://doi.org/10.1016/j.nanoen.2021.106171.

Johnsen, S., He, J.Q., Androulakis, J., Dravid, V.P., Todorov, I., Chung, D.Y., Kanatzidis, M.G. (2011). Nanostructures boost the thermoelectric performance of PbS. *J. Am. Chem. Soc.*, 133(10), 3460–3470. https://doi.org/10.1021/ja109138p.

Jood, P. and Ohta, M. (2015). Hierarchical architecturing for layered thermoelectric sulfides and chalcogenides. *Materials (Basel)*, 8(9), 6482–6483. https://doi.org/10.3390/ma8095315.

Jood, P., Ohta, M., Nishiate, H., Yamamoto, A., Lebedev, O.I., Berthebaud, D., Suekuni, K., Kunii, M. (2014). Microstructural control and thermoelectric properties of misfit layered sulfides $(LaS)_{1+m}TS_2$ (T = Cr, Nb): The natural superlattice systems. *Chem. Mater.*, 26(8), 2684–2692. https://doi.org/10.1021/cm5004559.

Jood, P., Ohta, M., Lebedev, O.I., Berthebaud, D. (2015). Nanostructural and microstructural ordering and thermoelectric property tuning in misfit layered sulfide $[(LaS)_x]_{1.14}NbS_2$. *Chem. Mater.*, 27(22), 7719–7728. https://doi.org/10.1021/acs.chemmater.5b03365.

Jood, P., Chetty, R., Ohta, M. (2020). Structural stability enables high thermoelectric performance in room temperature Ag_2Se. *J. Mater. Chem. A*, 8(26), 13024–13037. https://doi.org/10.1039/D0TA02614J.

Kanatzidis, M.G. (2005). Structural evolution and phase homologies for "Design" and prediction of solid-state compounds. *Acc. Chem. Res.*, 38(4), 359–368. https://doi.org/10.1021/ar040176w.

Kashida, S., Watanabe, N., Hasegawa, T., Iida, H., Mori, M., Savrasov, S. (2003). Electronic structure of Ag_2S, band calculation and photoelectron spectroscopy. *Solid State Ionics*, 158(1–2), 167–175. https://doi.org/https://doi.org/10.1016/S0167-2738(02)00768-3.

Katsuyama, S., Tanaka, Y., Hashimoto, H., Majima, K., Nagai, H. (1997). Effect of substitution of la by alkaline earth metal on the thermoelectric properties and the phase stability of γ-La_3S_4. *J. Appl. Phys.*, 82(11), 5513–5519. https://doi.org/10.1063/1.366409.

Kikuchi, Y., Bouyrie, Y., Ohta, M., Suekuni, K., Aihara, M., Takabatake, T. (2016). Vanadium-free colusites $Cu_{26}A_2Sn_6S_{32}$ (A = Nb, Ta) for environmentally friendly thermoelectrics. *J. Mater. Chem. A*, 4(39), 15207–15214. https://doi.org/10.1039/c6ta05945g.

Korkosz, R.J., Chasapis, T.C., Lo, S.-H., Doak, J.W., Kim, Y.J., Wu, C.-I., Hatzikraniotis, E., Hogan, T.P., Seidman, D.N., Wolverton, C. et al. (2014). High ZT in p-type $(PbTe)_{1-2x}(PbSe)_x(PbS)_x$ Thermoelectric materials. *J. Am. Chem. Soc.*, 136(8), 3225–3237. https://doi.org/10.1021/ja4121583.

Koyano, M., Negishi, H., Ueda, Y., Sasaki, M., Inoue, M. (1986). Electrical resistivity and thermoelectric power of intercalation compounds M_xTiS_2 (M = Mn, Fe, Co, and Ni). *Phys. Stat. Sol.*, 138(1), 357–363. https://doi.org/10.1002/pssb.2221380137.

Kumar, V.P., Supka, A.R., Lemoine, P., Lebedev, O.I., Raveau, B., Suekuni, K., Nassif, V., Al Orabi, R.A., Fornari, M., Guilmeau, E. (2019). High power factors of thermoelectric colusites $Cu_{26}T_2Ge_6S_{32}$ (T = Cr, Mo, W): Toward functionalization of the conductive "Cu–S" network. *Adv. Energy Mater.*, 9(6), 1803249, 1–11. https://doi.org/10.1002/aenm.201803249.

Lefèvre, R., Berthebaud, D., Mychinko, M.Y., Lebedev, O.I., Mori, T., Gascoin, F., Maignan, A. (2016). Thermoelectric properties of the chalcopyrite $Cu_{1-x}M_x FeS_{2-y}$ series (M = Mn, Co, Ni). *RSC Adv.*, 6(60), 55117–55124. https://doi.org/10.1039/c6ra10046e.

Li, J.H., Tan, Q., Li, J.F. (2013). Synthesis and property evaluation of $CuFeS_{2-x}$ as earth-abundant and environmentally-friendly thermoelectric materials. *J. Alloys Compd.*, 551, 143–149. https://doi.org/10.1016/j.jallcom.2012.09.067.

Li, Y.L., Zhang, T.S., Qin, Y.T., Day, T., Snyder, G.J., Shi, X., Chen, L.D. (2014). Thermoelectric transport properties of diamond-like $Cu_{1-x}Fe_{1+x}S_2$ tetrahedral compounds. *J. Appl. Phys.*, 116, 203705, 1–8. https://doi.org/10.1063/1.4902849.

Liang, J.S., Wang, T., Qiu, P.F., Yang, S.Q., Ming, C., Chen, H.Y., Song, Q.F., Zhao, K.P., Wei, T.R., Ren, D.D. et al. (2019). Flexible thermoelectrics: From silver chalcogenides to full-inorganic devices. *Energy Environ. Sci.*, 12(10), 2983–2990. https://doi.org/10.1039/C9EE01777A.

Liang, J.S., Qiu, P.F., Zhu, Y., Huang, H., Gao, Z.Q., Zhang, Z., Shi, X., Chen, L.D. (2020). Crystalline structure-dependent mechanical and thermoelectric performance in $Ag_2S_{1-x}S_x$. *Research*, 6591981, 1–10. https://doi.org/10.34133/2020/6591981.

Liu, M.L., Huang, F.Q., Chen, L.D., Wang, Y.M., Wang, Y.H., Li, G.F., Zhang, Q. (2007a). *p*-type transparent conductor: Zn-doped $CuAlS_2$. *Appl. Phys. Lett.*, 90(7), 16–19. https://doi.org/10.1063/1.2591415.

Liu, M.L., Wang, Y.M., Huang, F.Q., Chen, L.D., Wang, W.D. (2007b). Optical and electrical properties study on p-type conducting CuAlS$_{2+x}$ with wide band gap. *Scr. Mater.*, 57(12), 1133–1136. https://doi.org/10.1016/j.scriptamat.2007.08.015.

Liu, H.L., Shi, X., Xu, F.F., Zhang, L.L., Zhang, W.Q., Chen, L.D., Li, Q., Uher, C., Day, T., Snyder, G.J. (2012). Copper Ion liquid-like thermoelectrics. *Nat. Mater*, 11(5), 422–425. https://doi.org/10.1038/nmat3273.

Logothetis, E.M., Kaiser, W.J., Kukkonen, C.A., Faile, S.P., Colella, R., Gambold, J. (1980). Transport properties and the semiconducting nature of TiS$_2$. *Phys. B+C*, 99(1–4), 193–198. https://doi.org/10.1016/0378-4363(80)90231-4.

Lu, X., Morelli, D.T., Xia, Y., Zhou, F., Ozolins, V., Chi, H., Zhou, X.Y., Uher, C. (2013). High performance thermoelectricity in earth-abundant compounds based on natural mineral tetrahedrites. *Adv. Energy Mater.*, 3(3), 342–348. https://doi.org/10.1002/aenm.201200650.

Lu, X., Morelli, D.T., Xia, Y., Ozolins, V. (2015). Increasing the thermoelectric figure of merit of tetrahedrites by co-Doping with nickel and zinc. *Chem. Mater.*, 27(2), 408–413. https://doi.org/10.1021/cm502570b.

Makovicky, E. (2006). Crystal structures of sulfides and other chalcogenides. *Rev. Mineral. Geochem*, 61(1), 7–125. https://doi.org/10.2138/rmg.2006.61.2.

Miyatami, S. (1968). α-Ag$_2$S as a mixed conductor. *J. Phys. Soc. Jpn.*, 24(2), 328–336. https://doi.org/10.1143/JPSJ.24.328.

Miyazaki, Y., Ogawa, H., Kajitani, T. (2004). Preparation and thermoelectric properties of misfit-layered sulfide [Yb$_{1.90}$S$_2$]$_{0.62}$NbS$_2$. *Jpn J. Appl. Phys.*, 43(9A/B), L1202–L1204. https://doi.org/10.1143/jjap.43.11202.

Miyazaki, Y., Ogawa, H., Nakajo, T., Kikuchii, Y., Hayashi, K. (2013). Crystal structure and thermoelectric properties of misfit-layered sulfides [Ln$_2$S$_2$]$_p$NbS$_2$ (Ln = lanthanides). *J. Electron. Mater.*, 42(7), 1335–1339. https://doi.org/10.1007/s11664-012-2443-5.

Mizoguchi, H., Hosono, H., Ueda, N., Kawazoe, H. (1995). Preparation and electrical properties of Bi$_2$S$_3$ whiskers. *Jpn. J. Appl. Phys.*, 78(2), 1376–1378. https://doi.org/10.1063/1.360315.

Mrotzek, A. and Kanatzidis, M.G. (2003a). "Design" in solid-state chemistry based on phase homologies. The concept of structural evolution and the new megaseries A$_m$[M$_{1+l}$Se$_{2+l}$]$_2$ $_m$[M$_{2l+n}$Se$_{2+3l+n}$]. *Acc. Chem. Res.*, 36(2), 111–119. https://doi.org/10.1021/ar020099+.

Mrotzek, A. and Kanatzidis, M.G. (2003b). Tropochemical cell-twinning in the new quaternary bismuth selenides K$_x$Sn$_{6-2x}$Bi$_{2+x}$Se$_9$ and KSn$_5$Bi$_5$Se$_{13}$. *Inorg. Chem.*, 42(22), 7200–7206. https://doi.org/10.1021/ic034252n.

Nakahara, J.F., Takeshita, T., Tschetter, M.J., Beaudry, B.J., Gschneidner, K.A. (1988). Thermoelectric properties of lanthanum sulfide with Sm, Eu, and Yb Additives. *J. Appl. Phys.*, 63(7), 2331–2336. https://doi.org/10.1063/1.341049.

Ohta, M. and Hirai, S. (2009). Thermoelectric properties of NdGd$_{1+x}$S$_3$ prepared by CS$_2$ sulfurization. *J. Electron. Mater.*, 38(7), 1287–1292. https://doi.org/10.1007/s11664-009-0660-3.

Ohta, M., Yuan, H.B., Hirai, S., Uemura, Y., Shimakage, K. (2004). Preparation of R$_2$S$_3$ (R : La, Pr, Nd, Sm) powders by sulfurization of oxide powders using CS$_2$ Gas. *J. Alloys Compd.*, 374(1–2), 112–115. https://doi.org/10.1016/j.jallcom.2003.11.081.

Ohta, M., Yuan, H.B., Hirai, S., Yajima, Y., Nishimura, T., Shimakage, K. (2008). Thermoelectric properties of Th$_3$P$_4$-type rare-earth sulfides Ln$_2$S$_3$ (Ln = Gd, Tb) prepared by reaction of their oxides with CS$_2$ gas. *J. Alloys Compd.*, 451(1–2), 627–631. https://doi.org/10.1016/j.jallcom.2007.04.078.

Ohta, M., Hirai, S., Kato, H., Sokolov, V.V., Bakovets, V.V. (2009a). Thermal decomposition of NH$_4$SCN for preparation of Ln$_2$S$_3$ (Ln = La and Gd) by sulfurization. *Mater. Trans.*, 50(7), 1885–1889. https://doi.org/10.2320/matertrans.M2009060.

Ohta, M., Kuzuya, T., Sasaki, H., Kawasaki, T., Hirai, S. (2009b). Synthesis of multinary rare-earth sulfides PrGdS$_3$, NdGdS$_3$, and SmEuGdS$_4$, and investigation of their thermoelectric properties. *J. Alloys Compd.*, 484(1–2), 268–272. https://doi.org/10.1016/j.jallcom.2009.04.076.

Ohta, M., Obara, H., Yamamoto, A. (2009c). Preparation and thermoelectric properties of Chevrel-phase Cu$_x$Mo$_6$S$_8$ (2.0≤x≤4.0). *Mater. Trans.*, 50(9), 2129–2133. https://doi.org/10.2320/matertrans.MAW200918.

Ohta, M., Yamamoto, A., Obara, H. (2010). Thermoelectric properties of Chevrel-phase sulfides M$_x$Mo$_6$S$_8$ (M: Cr, Mn, Fe, Ni). *J. Electron. Mater.*, 39(9), 2117–2121. https://doi.org/10.1007/s11664-009-0975-0.

Ohta, M., Hirai, S., Kuzuya, T. (2011). Preparation and thermoelectric properties of LaGd$_{1+x}$S$_3$ and SmGd$_{1+x}$S$_3$. *J. Electron. Mater.*, 40(5), 537–542. https://doi.org/10.1007/s11664-010-1436-5.

Ohta, M., Satoh, S., Kuzuya, T., Hirai, S., Kunii, M., Yamamoto, A. (2012). Thermoelectric properties of Ti$_{1+x}$S$_2$ prepared by CS$_2$ sulfurization. *Acta Mater.*, 60(20), 7232–7240. https://doi.org/10.1016/j.actamat.2012.09.035.

Ohta, M., Chung, D.Y., Kunii, M., Kanatzidis, M.G. (2014). Low lattice thermal conductivity in Pb$_5$Bi$_6$Se$_{14}$, Pb$_3$Bi$_2$S$_6$, and PbBi$_2$S$_4$: Promising thermoelectric materials in the cannizzarite, lillianite, and galenobismuthite homologous series. *J. Mater. Chem. A*, 2(47), 20048–20058. https://doi.org/10.1039/C4TA05135A.

Perrin, A., Perrin, C., Chevrel, R. (2019). Chevrel phases: Genesis and developments, structure and bonding. In *Ligated Transition Metal Clusters in Solid-state Chemistry*, Volume 180. Springer, Cham. https://doi.org/10.1007/430_2019_35.

Powell, A.V. (2019). Recent developments in earth-abundant copper-sulfide thermoelectric materials. *J. Appl. Phys.*, 126(10), 100901, 1–23. https://doi.org/10.1063/1.5119345.

Putri, Y.E., Wan, C.L., Zhang, R.Z., Mori, T., Koumoto, K. (2013). Thermoelectric performance enhancement of $(BiS)_{1.2}(TiS_2)_2$ misfit layer sulfide by chromium doping. *J. Adv. Ceram.*, 2(1), 42–48. https://doi.org/10.1007/s40145-013-0040-6.

Qiu, P.F., Agne, M.T., Liu, Y.Y., Zhu, Y.Q., Chen, H.Y., Mao, T., Yang, J., Zhang, W.Q., Haile, S.M., Zeier, W.G. et al. (2018). Suppression of atom motion and metal deposition in mixed ionic electronic conductors. *Nat. Commun.*, 9(1), 4–11. https://doi.org/10.1038/s41467-018-05248-8.

Qiu, P.F., Mao, T., Huang, Z.F., Xia, X.G., Liao, J.C., Agne, M.T., Gu, M., Zhang, Q.H., Ren, D.D., Bai, S.Q. et al. (2019). High-efficiency and stable thermoelectric module based on liquid-like materials. *Joule*, 3(6), 1538–1548. https://doi.org/10.1016/j.joule.2019.04.010.

Qin, Y.X., Hong, T., Qin, B.C., Wang, D.Y., He, W.K., Gao, X., Xiao, Y., Zhao, L.D. (2021). Contrasting Cu roles lead to high ranged thermoelectric performance of PbS. *Adv. Funct. Mater.*, 31(34), 2102185, 1–9. https://doi.org/10.1002/adfm.202102185.

Roche, C., Pecheur, P., Toussaint, G., Jenny, A., Scherrer, H., Scherrer, S. (1998). Study of Chevrel phases for thermoelectric applications: Band structure calculations on compounds (M = Metal). *J. Phys. Condens. Matter*, 10(21), L333–L339. https://doi.org/10.1088/0953-8984/10/21/001.

Scanlon, W.W. (1959). Recent advances in the optical and electronic properties of PbS, PbSe, PbTe and their alloys. *J. Phys. Chem. Solids*, 8, 423–428. https://doi.org/10.1016/0022-3697(59)90379-8.

Shi, X., Chen, H.Y., Hao, F., Liu, R.H., Wang, T., Qiu, P.F., Burkhardt, U., Grin, Y., Chen, L.D. (2018). Room-temperature ductile inorganic semiconductor. *Nat. Mater.*, 17(5), 421–426. https://doi.org/10.1038/s41563-018-0047-z.

Shimizu, Y., Suekuni, K., Saito, H., Lemoine, P., Guilmeau, E., Raveau, B., Chetty, R., Ohta, M., Takabatake, T., Ohtaki, M. (2021). Synergistic effect of chemical substitution and insertion on the thermoelectric performance of $Cu_{26}V_2Ge_6S_{32}$ colusite. *Inorg. Chem.*, 60(15), 11364–11373. https://doi.org/10.1021/acs.inorgchem.1c01321.

Shojaei, M., Shokuhfar, A., Zolriasatein, A., Ostovari Moghaddam, A. (2022). Enhanced thermoelectric performance of CuAlS$_2$ by adding multi-walled carbon nanotubes. *Adv. Powder Technol.*, 33(2), 103445, 1–10. https://doi.org/10.1016/j.apt.2022.103445.

Slack, G.A. (1995). New materials and performance limits for thermoelectric cooling. In *CRC Handbook of Thermoelectrics*, Rowe, D.M. (ed.). CRC Press, Boca Raton.

Snyder, G.J. and Toberer, E.S. (2008). Complex thermoelectric materials. *Nat. Mater.*, 7(2), 105–114. https://doi.org/10.1038/nmat2090.

Sotnikov, A.V., Jood, P., Ohta, M. (2020). Enhancing the thermoelectric properties of misfit layered sulfides (MS)$_{1.2+q}$(NbS$_2$)$_n$ (M = Gd and Dy) through structural evolution and compositional tuning. *ACS Omega*, 5(22), 13006–13013. https://doi.org/10.1021/acsomega.0c00908.

Suekuni, K. and Takabatake, T. (2016). Research update: Cu–S based synthetic minerals as efficient thermoelectric materials at medium temperatures. *APL Mater.*, 4(10), 104503. https://doi.org/10.1063/1.4955398.

Suekuni, K., Tsuruta, K., Ariga, T., Koyano, M. (2012). Thermoelectric properties of mineral tetrahedrites Cu$_{10}$Tr$_2$Sb$_4$S$_{13}$ with low thermal conductivity. *Appl. Phys. Express*, 5(5), 051201, 1–3. https://doi.org/10.1143/APEX.5.051201.

Suekuni, K., Tsuruta, K., Kunii, M., Nishiate, H., Nishibori, E., Maki, S., Ohta, M., Yamamoto, A., Koyano, M. (2013). High-performance thermoelectric mineral Cu$_{12-x}$Ni$_x$Sb$_4$S$_{13}$ tetrahedrite. *J. Appl. Phys.*, 113(4), 043712, 1–5. https://doi.org/10.1063/1.4789389.

Suekuni, K., Kim, F.S., Nishiate, H., Ohta, M., Tanaka, H.I., Takabatake, T. (2014). High-performance thermoelectric minerals: Colusites Cu$_{26}$V$_2$M$_6$S$_{32}$ (M = Ge, Sn). *Appl. Phys. Lett.*, 105(13), 132107, 1–4. https://doi.org/10.1063/1.4896998.

Suekuni, K., Lee, C.H., Tanaka, H.I., Nishibori, E., Nakamura, A., Kasai, H., Mori, H., Usui, H., Ochi, M., Hasegawa, T. et al. (2018). Retreat from stress: Rattling in a planar coordination. *Adv. Mater.*, 30(13), 1706230, 1–6. https://doi.org/10.1002/adma.201706230.

Suekuni, K., Shimizu, Y., Nishibori, E., Kasai, H., Saito, H., Yoshimoto, D., Hashikuni, K., Bouyrie, Y., Chetty, R., Ohta, M. et al. (2019). Atomic-scale phonon scatterers in thermoelectric colusites with a tetrahedral framework structure. *J. Mater. Chem. A*, 7(1), 228–235. https://doi.org/10.1039/c8ta08248k.

Takeshita, T., Gschneidner, K.A., Beaudry, B.J. (1985). Preparation of γ-LaS$_y$ (1.33< y <1.50) alloys by the pressure-assisted reaction sintering method and their thermoelectric properties. *J. Appl. Phys.*, 57(10), 4633–4637. https://doi.org/10.1063/1.335373.

Tan, G.J., Ohta, M., Kanatzidis, M.G. (2019). Thermoelectric power generation: From new materials to devices. *Philos. Trans. R. Soc. A Math. Phys. Eng. Sci.*, 377(2152), 20180450, 1–28. https://doi.org/10.1098/rsta.2018.0450.

Tian, R.M., Wan, C.L., Wang, Y.F., Wei, Q.S., Ishida, T., Yamamoto, A., Tsuruta, A., Shin, W.S., Li, S., Koumoto, K. (2017). A solution-processed TiS_2/organic hybrid superlattice film towards flexible thermoelectric devices. *J. Mater. Chem. A*, 5(2), 564–570. https://doi.org/10.1039/C6TA08838D.

Toide, T., Utsunomiya, T., Sato, M., Hoshino, Y., Hatano, T., Akimoto, Y. (1973). Preparation of lanthanum sulfides using carbon disulfide as a sulfurizing agent and the change of these sulfides on heating in Air. *Bull. Tokyo Inst. Technol.*, 117, 41–48.

Touloukian, Y.S., Kirby, R.K., Taylor, R.E., Desai, P.D. (1975). *Thermophysical Propetties of Matter-The TPRC Data Series – Vol. 12. Thermal Exapnsion Metallic Elements and Alloys.* Plenum, New York.

Tsubota, T., Ohtaki, M., Eguchi, K. (1999). Thermoelectric properties of chevrel-type sulfides AMo_6S_8. (A = Fe, Ni, Ag, Zn, Sn, Pb, Cu). *J. Ceram. Soc. Jpn*, 107(1248), 697–701. https://doi.org/10.2109/jcersj.107.697.

Tsujii, N. and Mori, T. (2013). High thermoelectric power factor in a carrier-doped magnetic semiconductor $CuFeS_2$. *Appl. Phys. Express*, 6(4), 043001, 1–4. https://doi.org/10.7567/APEX.6.043001.

Tsujii, N., Mori, T., Isoda, Y. (2014). Phase stability and thermoelectric properties of $CuFeS_2$-Based magnetic semiconductor. *J. Electron. Mater*, 43(6), 2371–2375. https://doi.org/10.1007/s11664-014-3072-y.

Wan, C.L., Wang, Y.F., Wang, N., Koumoto, K. (2010). Low-thermal-conductivity $(MS)_{1+x}(TiS_2)_2$ (M = Pb, Bi, Sn) misfit layer compounds for bulk thermoelectric materials. *Materials (Basel)*, 3(4), 2606–2617. https://doi.org/10.3390/ma3042606.

Wan, C.L., Wang, Y.F., Wang, N., Norimatsu, W., Kusunoki, M., Koumoto, K. (2011). Intercalation: Building a natural superlattice for better thermoelectric performance in layered chalcogenides. *J. Electron. Mater.*, 40(5), 1271–1280. https://doi.org/10.1007/s11664-011-1565-5.

Wan, C.L., Wang, Y.F., Norimatsu, W., Kusunoki, M., Koumoto, K. (2012). Nanoscale stacking faults induced low thermal conductivity in thermoelectric layered metal sulfides. *Appl. Phys. Lett.*, 100(10), 101913, 1–4. https://doi.org/10.1063/1.3691887.

Wan, C.L., Gu, X.X., Dang, F., Itoh, T., Wang, Y.F., Sasaki, H., Kondo, M., Koga, K., Yabuki, K., Snyder, G.J. et al. (2015). Flexible n-type thermoelectric materials by organic intercalation of layered transition metal dichalcogenide TiS_2. *Nat. Mater.*, 14(6), 622–627. https://doi.org/10.1038/nmat4251.

Wang, Y.F., Pan, L., Li, C., Tian, R.M., Huang, R., Hu, X.H., Chen, C.C., Bao, N.Z., Koumoto, K., Lu, C.H. (2018). Doubling the *ZT* record of TiS$_2$-based thermoelectrics by incorporation of ionized impurity scattering. *J. Mater. Chem. C*, 6(35), 9345–9353. https://doi.org/10.1039/C8TC00914G.

Wang, H.T., Ma, H.Q., Duan, B., Geng, H.Y., Zhou, L., Li, J.L., Zhang, X.L., Yang, H.J., Li, G.D., Zhai, P.C. (2021a). High-pressure rapid preparation of high-performance binary silver sulfide thermoelectric materials. *ACS Appl. Energy Mater*, 4(2), 1610–1618. https://doi.org/10.1021/acsaem.0c02810.

Wang, Y.A., Pang, H., Guo, Q.S., Tsujii, N., Baba, T., Baba, T., Mori, T. (2021b). Flexible *n*-type abundant chalcopyrite/PEDOT:PSS/Graphene hybrid film for thermoelectric device utilizing low-grade heat. *ACS Appl. Mater. Interfaces*, 13(43), 51245–51254. https://doi.org/10.1021/acsami.1c15232.

Wiegers, G.A. (1996). Misfit layer compounds: Structures and physical properties. *Prog. Solid State Chem.*, 24(1–2), 1–139. https://doi.org/10.1016/0079-6786(95)00007-0.

Wood, C. (1988). Materials for thermoelectric energy conversion. *Rep. Prog. Phys.*, 51(4), 459–539. https://doi.org/10.1088/0034-4885/51/4/001.

Wood, C., Lockwood, A., Parker, J., Zoltan, A., Zoltan, D., Danielson, L.R., Raag, V. (1985). Thermoelectric properties of lanthanum sulfide. *J. Appl. Phys.*, 58(4), 1542–1547. https://doi.org/10.1063/1.336088.

Wu, H.J., Zhao, L.D., Zheng, F.S., Wu, D., Pei, Y.L., Tong, X., Kanatzidis, M.G., He, J.Q. (2014). Broad temperature plateau for thermoelectric figure of merit ZT>2 in phase-separated PbTe$_{0.7}$S$_{0.3}$. *Nat. Commun.*, 5(1), 4515, 1–9. https://doi.org/10.1038/ncomms5515.

Yan, Y.C., Wu, H., Wang, G.Y., Lu, X., Zhou, X.Y. (2018). High thermoelectric performance balanced by electrical and thermal transport in tetrahedrites Cu$_{12+x}$Sb$_4$S$_{12}$Se. *Energy Storage Mater.*, 13, 127–133. https://doi.org/10.1016/j.ensm.2018.01.006.

Yang, S.Q., Gao, Z.Q., Qiu, P.F., Liang, J.S., Wei, T.R., Deng, T.T., Xiao, J., Shi, X., Chen, L.D. (2021). Ductile Ag$_{20}$S$_7$Te$_3$ with excellent shape-conformability and high thermoelectric performance. *Adv. Mater.*, 33(10), 2007681, 1–9. https://doi.org/https://doi.org/10.1002/adma.202007681.

Yin, C., Hu, Q., Tang, M.J., Liu, H.T., Chen, Z.Y., Wang, Z.S., Ang, R. (2018). Boosting the thermoelectric performance of misfit-layered (SnS)$_{1.2}$(TiS$_2$)$_2$ by a Co- and Cu-substituted alloying effect. *J. Mater. Chem. A*, 6(45), 22909–22914. https://doi.org/10.1039/C8TA08426B.

Yuan, H.B., Kuzuya, T., Ohta, M., Hirai, S. (2010). Low-temperature formation of cubic Th$_3$P$_4$-Type gadolinium and holmium sesquisulfides. *J. MMIJ*, 126(7), 450–455. https://doi.org/10.2473/journalofmmij.126.450.

Zebarjadi, M., Esfarjani, K., Dresselhaus, M.S., Ren, Z.F., Chen, G. (2012). Perspectives on thermoelectrics: From fundamentals to device applications. *Energy Environ. Sci.*, 5(1), 5147–5162. https://doi.org/10.1039/C1EE02497C.

Zhang, Q.H., Huang, X.Y., Bai, S.Q., Shi, X., Uher, C., Chen, L.D. (2016). Thermoelectric devices for power generation: Recent progress and future challenges. *Adv. Eng. Mater.*, 18(2), 194–213. https://doi.org/10.1002/adem.201500333.

Zhao, L.D., Lo, S.H., He, J.Q., Li, H., Biswas, K., Androulakis, J., Wu, C.I., Hogan, T.P., Chung, D.Y., Dravid, V.P. et al. (2011). High performance thermoelectrics from earth-abundant materials: Enhanced figure of merit in PbS by second phase nanostructures. *J. Am. Chem. Soc.*, 133(50), 20476–20487. https://doi.org/10.1021/ja208658w.

Zhao, J., Hao, S.Q., Islam, S.M., Chen, H.J., Tan, G.J., Ma, S.L., Wolverton, C., Kanatzidis, M.G. (2019a). Six quaternary chalcogenides of the pavonite homologous series with ultralow lattice thermal conductivity. *Chem. Mater.*, 31(9), 3430–3439. https://doi.org/10.1021/acs.chemmater.9b00585.

Zhao, K., Qiu, P., Shi, X., Chen, L. (2019b). Recent advances in liquid-like thermoelectric materials. *Adv. Funct. Mater.*, 30(8), 1903867, 1–19. https://doi.org/10.1002/adfm.201903867.

Zhou, T., Lenoir, B., Colin, M., Dauscher, A., Al Orabi, R.A., Gougeon, P., Potel, M., Guilmeau, E. (2011). Promising thermoelectric properties in Ag$_x$Mo$_9$Se$_{11}$ compounds (3.4≤x≤3.9). *Appl. Phys. Lett.*, 98(16), 162106, 1–3. https://doi.org/10.1063/1.3579261.

Zhou, W.X., Wu, D., Xie, G.F., Chen, K.Q., Zhang, G. (2020). α-Ag$_2$S: A ductile thermoelectric material with high ZT. *ACS Omega*, 5(11), 5796–5804. https://doi.org/10.1021/acsomega.9b03929.

5

A Concise Review of Strongly Correlated Oxides

Ichiro TERASAKI
Department of Physics, Nagoya University, Japan

5.1. Introduction to electron correlation

Let us start from what differentiates between metals and insulators. This has been a central issue in solid state physics. At the beginning of the 20th century, we knew that salt and copper formed a crystal, i.e. a substance made from periodic arrangement of a vast number of atoms, but did not understand why salt was insulating and why copper was conductive. Now, we understand this on the basis of the band theory, as explained in all the introductory textbooks for solid state physics (Kittel et al. 1996).

We should note here that the band theory is based on the so-called one-electron approximation and does not properly take the electron–electron interaction into account. The conduction electrons are negatively charged, and they tend to keep away from one another as far as possible in a correlated manner in order to minimize the repulsive Coulomb interaction. If this correlation is strong enough, such metals begin to show insulating behavior, and predictions from the band theory are broken down. Such systems are called strongly correlated electron systems, which many

---

For a color version of all the figures in this chapter, see www.iste.co.uk/akinaga/thermoelectric1.zip.

Thermoelectric Micro/Nano Generators 1,
coordinated by Hiroyuki AKINAGA, Atsuko KOSUGA,
Takao MORI and Gustavo ARDILA.
© ISTE Ltd 2023.

transition-metal oxides belong to (Dagotto 2005). A prime example is the high-temperature superconducting copper oxides (Lee et al. 2006b), in which no band calculations have predicted superconductivity before.

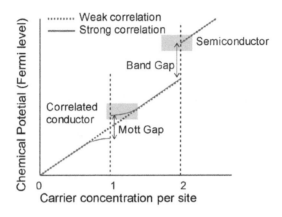

Figure 5.1. *Schematic of chemical potential plotted as a function of carrier concentration per site. The dotted line is a simplified curve for an uncorrelated (weakly correlated) electron system, while the solid curve corresponds to a strongly correlated electron system. We assume that the electronic states adiabatically transit from weakly correlated to strongly correlated ones*

We intuitively explain how the electron correlation modifies the electronic states. Figure 5.1 schematically shows the chemical potential μ plotted as a function of carrier concentration. Here, we suppose a simple solid which includes only one element responsible for the electrical conduction per unit cell, and we also assume no degeneracy in the valence band. In such cases, the carrier concentration can be expressed in terms of carrier number per site (n). When $n = 0$, the valence band is empty, and thus the system is insulating. When $n = 2$, the valence band is completely filled by the electrons with up and down spins, and there is a finite energy gap to the next valence band (also called the conduction band). The system is thus again insulating, which explains why solid helium is a tough insulator.

An interesting case is in $n = 1$, where the valence band is half-filled, and one electron stays in average at each site. In this situation, the electron correlation is most enhanced, and the system can be easily insulating when the electron correlations are sufficiently strong. As shown in Figure 5.1, by adiabatically increasing the strength of electron correlation, an energy gap opens at $n = 1$, which is referred to as the "Mott gap" (Imada et al. 1998).

This situation is well captured in the Hubbard Hamiltonian expressed by

$$H = t \sum_{i,j,\sigma} c_{i\sigma}^{\dagger} c_{j\sigma} + U \sum_i n_{i\uparrow} n_{i\downarrow}, \qquad [5.1]$$

where $c_{i\sigma}^{\dagger}$ and $c_{i\sigma}$ are, respectively, the creation and annihilation operators of an electron with spin σ (= ↑ or ↓) at the site i. $n_{i\sigma} = c_{i\sigma}^{\dagger} c_{i\sigma}$ is the number operator at the site i, and takes the eigenvalue of 0 or 1 reflecting the Pauli principle. The first term represents the transfer energy from the site j to i, describing the energy band with a width proportional to t, while the second term represents the strength of the Coulomb repulsion U at the site i, meaning that two electrons with opposite spin sitting at the same site increase the total energy by U.

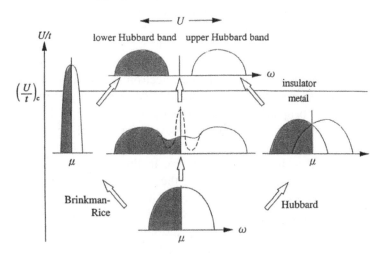

Figure 5.2. Schematic electronic states for the half-filling ($n = 1$) as a function of U/t (Fujimori et al. 2001)

Although the second term is a brave simplification of the Coulomb repulsion in the sense that the long-range potential of $|lr - lr'|^{-1}$ is approximated as the impulsive force of $\delta(lr - lr')$, the Hubbard Hamiltonian captures essential features of the strong correlation. Figure 5.2 schematically shows how the electronic states for $n = 1$ are modified as a function of U/t (Fujimori et al. 2001). At the bottom of the figure, a weakly correlated case is drawn, where the density of states (DOS) is half-filled with the chemical potential in the center of DOS. This is a typical picture of a metal, and the physical properties are determined by the electrons near the chemical potential (i.e. the Fermi energy). At the top of the figure, on the other hand, the large limit of U/t causes the Mott gap of U, and the lower (upper) Hubbard band is completely filled (empty). This is a typical picture of an insulator except for the nature of the Hubbard band; the lower (upper) Hubbard band accepts only one electron per site, whereas the conventional band accepts two. Thus, the Hubbard band is produced by many-body effects, and each one-electron state in the band cannot be expressed analytically.

In the intermediate state of U/t, the electronic states depend on approximation. Figure 5.2 shows three typical examples. As shown in the right of the figure, Hubbard treated the second term in equation [5.1] within a framework of mean-field approximation. The upper and lower Hubbard bands overlap, and the center of gravity for each DOS is separated by U. With increasing U, the two bands continuously separate to create the Mott gap. Brinkman and Rice treated the electron correlation through the Gutzwiller projection method, where the double occupancy at the same site is excluded by the projection operator $P = \Pi_i (1 - n_{i\uparrow} n_{i\downarrow})$. In this technique, the electrons move coherently in order to avoid the double occupancy at the same site, and eventually the effective mass is enhanced by the coherent motion, as shown in the left of the figure. In this situation, the bandwidth gets narrower with increasing U/t, and the effective mass enhanced by U finally diverges to become an insulator, but it is difficult to create the Mott gap. Recently, dynamical mean field theory (DMFT) was developed, where the effect of U can be exactly taken into the self-energy of one-electron state in the infinite dimensions. The DMFT calculation gives the one-electron electronic state, as shown in the center of the figure, where the Brinkman–Rice-like coherent peak at μ has a Hubbard-band-like incoherent peak at $\pm U/2$ from μ. In this picture, the mass enhancement and the gap growth simultaneously occur.

We have investigated why and how the electron correlation improves the thermoelectric performance in some transition-metal oxides. As shown in Figure 5.1, the chemical potential becomes flat near the Mott gap, implying a macroscopic number of degeneracy near the gap (i.e. $dn/d\mu \rightarrow \infty$) and a large entropy per site. In terms of the Hubbard band, the lower Hubbard band is filled, but still has spin degrees of freedom at each site; one electron stays at each site, but its spin orientation is undetermined either up or down. Peterson and Shastry (2010) have proposed that the Kelvin formula well describes the thermopower of strongly correlated materials. The Kelvin formula is given by

$$S_K = -\frac{1}{e}\left(\frac{\partial \mu}{\partial T}\right)_{N,V}, \qquad [5.2]$$

where N is the electron number, and V is the volume. The Gibbs–Duhem equation gives

$$-\frac{\partial \mu}{\partial T} = \left(\frac{\Sigma}{N}\right)_{N,P} \qquad [5.3]$$

where Σ is the entropy and P is the pressure. In solids far below the melting point, the constant-volume condition can be identified with the constant-pressure condition. Then, we arrive at $eS_K = \Sigma/N$, meaning that the entropy per electron determines the thermopower. Accordingly, S can be enhanced by the large entropy/degeneracy near $n = 1$ in strongly correlated electron systems. Here, we review the thermoelectric properties of transition-metal oxides, and explain the contribution of the strongly correlated electrons.

5.2. Electronic states of transition-metal oxides

Let us introduce possible energy levels of d orbitals of the M^{n+} ion surrounded by the six oxygen anions, as schematically drawn in Figure 5.3(a). When the Cartesian coordinate is taken as the directions to the oxygen ions, the fivefold degenerate d levels in vacuum split into doubly degenerate e_g level and triply degenerate t_{2g} level, as shown in Figure 5.3(b). This is due to the electrostatic Coulomb repulsion from the oxygen anions; the d orbitals extending away from the oxygen anions belong to the lower t_{2g} level, while the d orbitals pointing to the oxygen anions belong to the higher e_g level.

This electrostatic potential referred to as the ligand field competes with Hund's rule. Figure 5.3(c) shows how the d electrons occupy the energy levels. When the number of d electrons is from one to three (d^1, d^2, d^3), they occupy the t_{2g} levels with parallel spin configuration in order to satisfy Hund's rule. Similarly, when the number of d electrons is from eight to nine (d^8, d^9), holes occupy the e_g level with parallel spin configuration, instead of electrons.

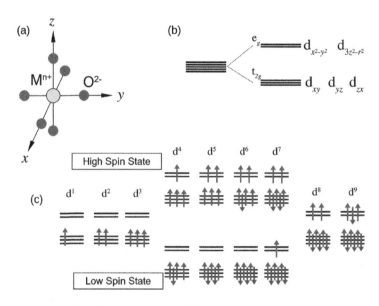

Figure 5.3. *(a) Transition-metal ion M^{n+} octahedrally coordinated by six oxygen anions. (b) The fivefold degenerate d orbitals and the energy levels of t_{2g} and e_g by the ligand field. (c) The spin state configuration for d^n electrons*

For the number of d electrons from four to seven (d^4, d^5, d^6, d^7), two different configurations are possible. When the ligand field is dominant, the d electrons occupy the lower t_{2g} level first. In contrast, when Hund's rule is dominant, the d electrons occupy t_{2g} and e_g levels in order to maximize the total spin number. The former is called the low-spin state, while the latter is called the high-spin state (Sugano et al. 1970).

5.3. 3D transition-metal oxides

The perovskite-type structure is widely seen in ternary transition-metal oxides, where the A cation forms a simple cubic lattice, the oxygen anion sits at the face-center site and the B cation occupies the body-center position, as schematically drawn in Figure 5.4(a). The A site usually accommodates large-size cations such as alkaline earth elements (Ca, Sr, Ba) and rare-earth element (Y, La, Ce, Pr, ..., Lu). In contrast, the B site usually accommodates small-sized cations such as transition metals (Ramadass 1978). Mostly, the A site controls the lattice volume and the formal valence of the B site, whereas the B site and the BO_6 octahedron determine the conductivity and magnetism.

Figure 5.4(b) schematically shows a phase diagram of transition-metal oxides with the perovskite structure ABO_3 (based on Fujimori (1992)). The horizontal axis schematically represents U/t. We can see that the vertical lines of d^0 (e.g. $SrTiO_3$) and d^1 (e.g. $LaTiO_3$) correspond to the band insulator and Mott insulator, respectively. The area between d^0 and d^1 corresponds to solid solutions. For example, only $x \sim 1$ makes the system insulating for $Sr_{1-x}La_xTiO_3$ (Tokura et al. 1993), while all the samples are insulating in $Ca_{1-x}Y_xTiO_3$ (Kumagai et al. 1993). The vertical line of d^n ($n \geq 4$) is named the "charge-transfer insulator" (Zaanen et al. 1985), which can be broadly interpreted as the "Mott insulator". The insulating area extends to lower near the vertical lines of d^n ($n \geq 1$), and conversely the metallic area mostly extends to upper near $d^{n-1/2}$ (quarter filling). Of course, the conductivity strongly depends on the species of the B-site cation; Cr, Mn and Fe oxides are mostly insulating, whereas Ti, Co, Ni and Cu oxides often become conductive by doping carriers (Imada et al. 1998).

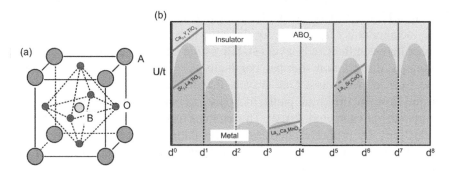

Figure 5.4. *(a) Crystal structure of simple cubic perovskite oxide ABO_3. (b) Schematic phase diagram for metal and insulator in transition-metal oxides with the perovskite-type structure (based on Fujimori (1992)). The vertical axis schematically represents the strength of correlation U/t*

5.3.1. Co oxides

Terasaki et al. (1997) discovered good thermoelectric properties in a single-crystal sample of $NaCo_2O_4$ (now known as Na_xCoO_2). This material showed a low resistivity of 200 μΩcm with a large thermopower of 100 μV/K at 300 K. Later, Fujita et al. (2001) measured high-temperature thermoelectric properties of their single-crystal samples, and found that the dimensionless figure of merit zT exceeded unity above 800 K. Since then, many related cobalt oxides have been synthesized and identified (Funahashi et al. 2000; Itoh and Terasaki 2000; Masset et al. 2000; Miyazaki et al. 2000; Maignan et al. 2002b; Pelloquin et al. 2002; Ishiwata et al. 2006; Kobayashi and Terasaki 2006; Yamauchi et al. 2011), some of which are found to be as good as Na_xCoO_2 (Funahashi and Shikano 2002; Shikano and Funahashi 2003). The record data are collected in Figure 5.5(a) (Prasad and Bhame 2020). Most data are concentrated in the cobalt oxides before 2010, where the single crystals show $zT > 0.9$ and even polycrystalline samples show $zT \sim 0.8$. Before the discovery of the layered cobalt oxide, oxides were not that competitive; people believed that the oxides were electrically poor but thermally good conductors. The appearance of the cobalt oxides has overturned the consensus in the community (Karppinen et al. 2002; Maignan et al. 2002a, 2002b; Koumoto et al. 2006; Hébert et al. 2007; Koumoto et al. 2013). After 2010, BiCuSeO-base compounds continued to update the record, which will be discussed in the next section.

The crystal structures of the layered cobalt oxides are schematically shown in Figure 5.5(b). The CdI_2-type CoO_2 layer consisting of edge-shared CoO_2 octahedra is responsible for the large power factor, and the block layer located between the CoO_2 layers stabilizes the structure electrostatically, and dominates the phonon thermal conductivity. The layered structure works quite well in two ways. One way is that the electric current and thermal current flow in different paths in space, as shown in the right panel of Figure 5.5(b) (Tada et al. 2010). This enables us to control the lattice thermal conductivity by properly choosing the insulating block layer (Satake et al. 2004; Takashima et al. 2021). This is a manifestation of electron-crystal and phonon-glass (Slack 1995; Takabatake et al. 2014).

The other way is that the CdI_2-type CoO_2 block enhances the power factor. Koshibae et al. (2000) extended the Heikes formula (Chaikin and Beni 1976) in order to include the spin and orbital degrees of freedom, and found the formula given by

$$S = \frac{k_B}{e} \log \frac{g_A}{g_B} \frac{x}{1-x}, \qquad [5.4]$$

where g_A and g_B are the degeneracies of the A and B ions, respectively, and x is the content of the A ions. As schematically drawn in Figure 5.3(c), the six electrons in the low-spin Co^{3+} ion (d^6) fully occupy the t_{2g} levels, so that degeneracy is unity. In contrast, in the low-spin Co^{4+} ion (d^5), one electron is removed out of the six electrons, and thus six states are degenerate (a spin degeneracy of two and an orbital degeneracy of three). Substituting $g_A = 6$ and $g_B = 1$ in equation [5.4], we evaluate the thermopower to be $k_B \log 6/e = 150$ μV/K in addition to the x-dependent term. This $\log g_A/g_B$ corresponds to a large entropy per electron, enhancing the thermopower through equations [5.2] and [5.3].

The relationship of the spin state of the Co^{3+} ions to the thermopower has been experimentally verified in the room-temperature ferromagnet $Sr_3YCo_4O_{10.5}$ (Kobayashi et al. 2005; Ishiwata et al. 2007; Nakao et al. 2011), where the spin state is easily controlled by pressure and a magnetic field (Matsunaga et al. 2010). Takahashi et al. (2018) found that the pressure changes the magnetization and thermopower in a consistent way. *Such spin and orbital degrees of freedom are absent in conventional thermoelectric materials, indicating the reason why the cobalt oxides possess extraordinarily high thermoelectric performance among oxides.*

This large entropy is evidenced by the specific heat measurement (Ando et al. 1999), and low-temperature thermopower (Wang et al. 2003; Lee et al. 2006a). Terasaki (2001, 2003) compared the thermodynamic properties of Na_xCoO_2 with those of heavy fermion intermetallics, in the sense that the spin/orbital degrees of freedom attach the conduction electrons. Behnia et al. (2004) proposed that the temperature-linear thermopower can be associated with the electron specific heat, and the origin of the large thermopower is interrelated between the cobalt oxides and the heavy fermion compounds.

The Heikes-formula-based argument neglects the itinerancy of the conduction electron. Koshibae and Maekawa (2001, 2003) theoretically studied the metallic conduction in the multiband Hubbard model, and discussed the relationship to the large thermopower. As thermoelectric materials, the layered cobalt oxides show good electrical conduction, and the band-picture should work well. At an early stage of the thermoelectric study in Na_xCoO_2, Singh (2000) already pointed out that the thermopower and specific heat of Na_xCoO_2 can be quantitatively understood from band calculation, and has suggested an importance of quantum fluctuation for good electrical conduction (Singh 2003). Kuroki and Arita (2007) suggested that a peculiar shape of the conduction band called the "pudding mold" can enhance the thermopower. Peterson and Shastry (2010) have shown by using the Kelvin formula that the hexagonal-lattice Hubbard model explains the thermopower of Na_xCoO_2 well.

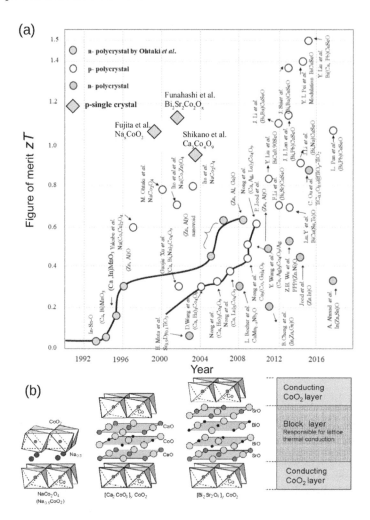

Figure 5.5. (a) The year dependence of the dimensionless figure of merit zT for oxide materials. The original figure is taken from Prasad and Bhame (2020), on which single-crystal data are added (Fujita et al. 2001; Funahashi and Shikano 2002; Shikano and Funahashi 2003). (b) The schematic of the layered cobalt oxides

5.3.2. Cu oxides

Although high-temperature superconducting cuprates (HTSC) are a prototypical example for strongly correlated oxides, they do not exhibit thermoelectric performance as good as other thermoelectric materials. This is partially because the electron configuration of copper in HTSC is close to d^9, in which the spin degrees of freedom are minimum ($S = 1/2$) with no orbital degrees of freedom. Thus, according to equation [5.4], large thermopower cannot be expected.

The thermopower and resistivity of heavily underdoped HTSC are nearly the same values as those of the thermoelectric cobalt oxides below room temperature (Horiuchi et al. 2010). However, the thermopower of the layered cobalt oxides continues to increase up to 1000 K, while the thermopower of heavily underdoped HTSC decreases with temperature (Takeda et al. 1994). The two-dimensional square lattice of the CuO_2 plane in HTSC is a bipartite with an electron–hole symmetry in the lowest order. As a result, the electron- and hole-like carriers are thermally excited equally to cancel the thermopower. In contrast, the triangular lattice in the CoO_2 plane has no such electron–hole symmetry, and a large thermopower persists at high temperatures. This remains valid in the presence of strong correlation, as pointed out by Merino and Mckenzie (2000). We evaluate the power factor of HTSC to be no more than 15 $\mu W/cmK^2$ (Horiuchi et al. 2010). Using the thermal conductivity of 50 mW/cmK, we evaluate zT to be 0.09 at maximum at 300 K. This evaluation is consistent with the early survey of Macklin and Moseley (1990), and indeed most of the literature data do not exceed this value (Liu et al. 2009).

At the end of this section, we will briefly introduce the copper oxyselenide BiCuSeO that was recently discovered as a thermoelectric material (Zhao et al. 2014). This class of materials (LaAgSO) was originally synthesized by Palazzi et al. (1980), and later Kamihara et al. (2006, 2008) discovered high-temperature superconductors now known as the "1111" phase. A possibility of thermoelectric material was first suggested by Zhao et al. (2010), where the 15-% Sr substitution improved zT up to 0.76 at 873 K.

This material is also a layered material in which the Bi_2O_2 layer and the Cu_2Se_2 layer alternately stack along the c axis. According to the band calculation, this material is a multiband semiconductor, whose valence bands consist of hybridized Cu 3D-Se4p and non-bonding Cu 3D. In the sense of electronic states, this material does not belong to Cu oxide, but is often introduced as a part of oxide thermoelectric materials. The electrons in this material are unlikely to be strongly correlated. Carrier doping is well described by the rigid-band picture, and the details of the Fermi surface are consistent between theory and experiment (Ren et al. 2019). As a thermoelectric material, the most advantageous feature is the low thermal conductivity coming from van-der-Waals bonding between the layers. More details can be seen in (Zhao et al. 2014) and the references therein.

5.3.3. Other 3D transition-metal oxides

We briefly mention other transition-metal oxides. The perovskite titanate $SrTiO_3$ has been extensively investigated as an n-type thermoelectric material. A small amount of Nb substitution for Ti or La substitution for Sr makes a large power factor arising from heavy effective mass (Okuda et al. 2001; Ohta et al. 2005). Ohta et al. (2007) found extraordinarily large thermopower in $SrTiO_3/Sr(Ti,Nb)O_3$ superlattices, and proposed that the two-dimensional electron gas enhances the thermoelectric performance. This material basically belongs to a conventional degenerate semiconductor, and not to correlated oxides. The Mott insulator $LaTiO_3$ is not a good thermoelectric material. Pálsson and Kotliar (1998) have pointed out that a simple Mott transition in the titanium oxide cannot improve thermoelectric properties. ZnO has been studied as an n-type thermoelectric material (Ohtaki et al. 1996), and shows a reasonably large $zT \sim 04$–0.6 at high temperatures (see Figure 5.5(a)). This is a conventional high-mobility semiconductor, where the electron correlation is weak. In the semiconducting manganese oxide $CaMnO_3$, the Mn ion has the electron configuration d^3, which filled t_{2g} levels with up spin. This situation practically works as a semiconductor with the energy gap between t_{2g} and e_g, where an electron is doped by the partial substitution of Nb for Mn (Thiel et al. 2015).

5.4. 4D transition-metal oxides

As is well known, the 3D orbital is localized in space, which often makes an insulating ground state. Since the orbital becomes broader in going from 3D to 5D, the 4D or 5D transition-metal oxides are more likely to be conductive. In particular, the 4D transition-metal oxide still has strong electron correlation, in which highly correlated metallic states are expected.

5.4.1. Rh oxides

Rhodium is located below cobalt in the periodic table, and thus is expected to have similar chemical properties. In fact, many cobalt oxides have their isomorphic rhodium oxides, and similar transport properties are reported in such layered rhodium oxides such as $Sr_xRh_2O_4$ (Okamoto et al. 2006), Bi-Ba/Sr-Rh-O (Okada and Terasaki 2005; Okada et al. 2005; Klein et al. 2006b; Kobayashi et al. 2007), K_xRhO_2 (Shibasaki et al. 2010; Okazaki et al. 2011) and $CuRhO_2$ (Shibasaki et al. 2006; Maignan et al. 2009). An important difference from the Co^{3+} ions is that the Rh^{3+} ions are stable in the low-spin state at all the temperatures of interest. This is advantageous to large thermopower based on equation [5.4]. The thermopower of $La_{1-x}Sr_xRhO_3$ (Ref (Terasaki et al. 2010)) makes a remarkable contrast to that of $La_{1-x}Sr_xCoO_3$ (Ref (Androulakis et al. 2004; Iwasaki et al. 2008)). The thermopower of $La_{1-x}Sr_xRhO_3$ continues to increase up to 800 K for all x, and remains larger than 150 μV/K at 800 K, as already discussed with equation [5.4], whereas the thermopower of $La_{1-x}Sr_xCoO_3$ rapidly decreases down to a few μV/K above 500 K due to the spin-state crossover (Iwasaki et al. 2008). Usui et al. (2009) have calculated the large thermopower in $LaRhO_3$ and $CuRhO_2$.

An intriguing question lies in the solid solution of cobalt and rhodium oxides. Figure 5.6(a) shows the resistivity of the perovskite oxide $La_{0.8}Sr_{0.2}Co_{1-x}Rh_xO_3$ (Shibasaki et al. 2011). The 20-% Sr substitution supplies sufficiently high carrier concentration in this system, making the samples metallic at room temperature. For $x = 0$, the hole in the Co^{4+} ion shows good conduction through the double exchange mechanism to the Co^{3+} ion in the intermediate spin state of $t_{2g}^5 e_g^1$. With increasing x, the Co^{3+} ions tend to stay at the low-spin state, and the resistivity systematically increases. The $x = 1$ sample shows almost temperature-independent resistivity. With decreasing temperature, the samples for $0 < x < 1$ show an upturn, and the holes are likely localized at low temperatures.

More interesting behavior is seen in the thermopower shown in Figure 5.6(b). The $x = 0$ sample shows small but metallic thermopower, which can be understood by applying equation [5.4] to the Co^{3+} ions in the intermediate-spin state. In contrast, in the $x = 1$ sample, the thermopower is as large as 80 μV/K at room temperature, and decreases almost linearly with decreasing temperature. This resembles the thermopower of Na_xCoO_2, implying the Rh^{3+} ions in the low-spin state. All the samples for $0 < x < 1$ show larger thermopower than $x = 0$ and 1, meaning that the thermopower cannot be understood by the average of $x = 0$ and 1. We have proposed that the spin-state disorder carries additional entropy to enhance the thermopower. We also suggest that this spin-state disorder is responsible for the non-metallic resistivity at low temperatures.

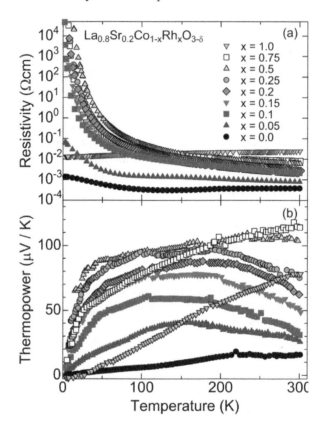

Figure 5.6. *(a) Resistivity and (b) thermopower of $La_{0.8}Sr_{0.2}Co_{1-x}Rh_xO_3$ (Shibasaki et al. 2011)*

Solid solutions of cobalt and rhodium oxides affect the spin states of Co^{3+}. Asai et al. (2011) found a weakly ferromagnetic state in $LaCo_{1-x}Rh_xO_3$, below 20 K, and suggested that the substituted Rh ions allow the neighboring Co^{3+} ions to be in the high-spin state. This speculation was checked by lattice expansion (Asai et al. 2013), infrared spectroscopy (Terasaki et al. 2017) and X-ray absorption (Sudayama et al. 2017). Theoretically, Knížek et al. (2012) have proposed that the substituted rhodium ion elastically and electrostatically stabilizes the neighboring cobalt ions in the high-spin state.

5.4.2. *Ru oxides*

A cubic perovskite ruthenate $SrRuO_3$ is an itinerant ferromagnet below 160 K, known as a bad metal whose resistivity continues to increase at high temperature due to strong correlation beyond the Ioffe–Regel limit (the mean free path cannot be shorter than the lattice parameters) (Klein et al. 1996). Klein et al. (2006a) found that the thermopower (~ 35 μV/K at 300 K) of $SrRuO_3$ is insensitive to impurity doping, and discussed this in terms of the modified Heikes formula of equation [5.4]. Hébert et al. (2015) extensively studied cubic perovskite-related ruthenates. Yamamoto et al. (2017) have shown that the field-dependent part of the thermopower obeys the Kelvin formula in the weakly ferromagnetic $CaRu_{0.8}Sc_{0.2}O_3$.

A layered perovskite ruthenate Ca_2RuO_4 crystallizes in the K_2NiF_4-type structure, where the conductive RuO_2 planes consisting of the corner-shared RuO_6 octahedra are stacked with the CaO double layer along the c axis. Nakatsuji et al. (1997) first synthesized this oxide, and later Alexander et al. (1999) found a first-order metal–insulator transition at approximately 360 K. We have studied the nonlinear conduction in Ca_2RuO_4 (Okazaki et al. 2013, 2014, 2020; Nishina et al. 2017; Tanabe et al. 2017), since Nakamura et al. (2013) discovered a giant nonlinear conduction a room temperature. Okazaki et al. (2013) established a measurement technique of nonlinear conduction using an infrared thermometer, and successfully separated the intrinsic nonlinear conduction from self-heating. Nishina et al. (2017) observed the Seebeck coefficient enhanced by a constant current flow. Very recently, Kawasaki et al. (2021) found that the in-plane thermal diffusivity of Ca_2RuO_4 decreases by 40% for a current density of 14 A/cm^2.

5.5. Concluding remarks

We have reviewed the current status of the thermoelectric properties of strongly correlated transition-metal oxides. Although they are mostly poor electric conductors, the dimensionless figure of merit is found to be close to unity at high temperatures. Taking the advantages in oxide materials (chemical stability at high temperatures, low mass density, easy processability, etc.) into account, Terasaki (2011) proposes some oxide materials as a good candidate for high-temperature thermoelectrics, as he previously pointed out that oxides are advantageous to the high-temperature thermoelectrics from the viewpoint of solid state physics of disordered materials.

5.6. References

Alexander, C., Cao, G., Dobrosavljevic, V., McCall, S., Crow, J., Lochner, E., Guertin, R. (1999). Destruction of the Mott insulating ground state of Ca_2RuO_4 by a structural transition. *Physical Review B*, 60, R8422–R8425.

Ando, Y., Miyamoto, N., Segawa, K., Kawata, T., Terasaki, I. (1999). Specific-heat evidence for strong electron correlations in the thermoelectric material $(Na, Ca)Co_2O_4$. *Physical Review B*, 60, 10580–10583.

Androulakis, J., Migiakis, P., Giapintzakis, J. (2004). $La_{0.95}Sr_{0.05}CoO_3$: An efficient room-temperature thermoelectric oxide. *Applied Physics Letters*, 84, 1099–1101.

Asai, S., Furuta, N., Yasui, Y., Terasaki, I. (2011). Weak ferromagnetism in $LaCo_{1-x}Rh_xO_3$: Anomalous magnetism emerging between two nonmagnetic end phases. *Journal of the Physical Society of Japan*, 80, 104705.

Asai, S., Okazaki, R., Terasaki, I., Yasui, Y., Kobayashi, W., Nakao, A., Kobayashi, K., Kumai, R., Nakao, H., Murakami, Y. (2013). Spin state of Co^{3+} in $LaCo_{1-x}Rh_xO_3$ investigated by structural phenomena. *Journal of the Physical Society of Japan*, 82, 114606.

Behnia, K., Jaccard, D., Flouquet, J. (2004). On the thermoelectricity of correlated electrons in the zero-temperature limit. *Journal of Physics: Condensed Matter*, 16, 5187–5198.

Chaikin, P. and Beni, G. (1976). Thermopower in the correlated hopping regime. *Physical Review B*, 13, 647–651.

Dagotto, E. (2005). Complexity in strongly correlated electronic systems. *Science*, 309, 257–262.

Fujimori, A. (1992). Electronic structure of metallic oxides: Band-gap closure and valence control. *Journal of Physics and Chemistry of Solids*, 53, 1595–1602.

Fujimori, A., Yoshida, T., Okazaki, K., Tsujioka, T., Kobayashi, K., Mizokawa, T., Onoda, M., Katsufuji, T., Taguchi, Y., Tokura, Y. (2001). Electronic structure of Mott–Hubbard-type transition-metal oxides. *Journal of Electron Spectroscopy and Related Phenomena*, 117, 277–286.

Fujita, K., Mochida, T., Nakamura, K. (2001). High-temperature thermoelectric properties of $Na_xCoO_{2-\delta}$ single crystals. *Japanese Journal of Applied Physics*, 40, 4644–4647.

Funahashi, R. and Shikano, M. (2002). $BiSr_2Co_2O_y$ whiskers with high thermoelectric figure of merit. *Applied Physics Letters*, 81, 1459–1461.

Funahashi, R., Matsubara, I., Ikuta, H., Takeuchi, T., Mizutani, U., Sodeoka, S. (2000). An oxide single crystal with high thermoelectric performance in air. *Japanese Journal of Applied Physics*, 39, L1127–L1129.

Hébert, S., Flahaut, D., Martin, C., Lemonnier, S., Noudem, J., Goupil, C., Maignan, A., Hejtmanek, J. (2007). Thermoelectric properties of perovskites: Sign change of the Seebeck coefficient and high temperature properties. *Progress in Solid State Chemistry*, 35, 457–467.

Hébert, S., Daou, R., Maignan, A. (2015). Thermopower in the quadruple perovskite ruthenates. *Physical Review B*, 91, 045106.

Horiuchi, Y., Tamura, W., Fujii, T., Terasaki, I. (2010). In-plane thermoelectric properties of heavily underdoped high-temperature superconductor $Bi_2Sr_2CaCu_2O_{8+\delta}$. *Superconductor Science and Technology*, 23, 065018.

Imada, M., Fujimori, A., Tokura, Y. (1998). Metal–insulator transitions. *Reviews of Modern Physics*, 70, 1039–1263.

Ishiwata, S., Terasaki, I., Kusano, Y., Takano, M. (2006). Transport properties of misfit-layered cobalt oxide $[Sr_2O_{2-\delta}]_{0.53}CoO_2$. *Journal of the Physical Society of Japan*, 75, 104716–104716.

Ishiwata, S., Kobayashi, W., Terasaki, I., Kato, K., Takata, M. (2007). Structure–property relationship in the ordered-perovskite-related oxide $Sr_{3.12}Er_{0.88}Co_4O_{10.5}$. *Physical Review B*, 75, 220406.

Itoh, T. and Terasaki, I. (2000). Thermoelectric properties of $Bi_{2.3-x}Pb_xSr_{2.6}Co_2O_y$ single crystals. *Japanese Journal of Applied Physics*, 39, 6658–6660.

Iwasaki, K., Ito, T., Nagasaki, T., Arita, Y., Yoshino, M., Matsui, T. (2008). Thermoelectric properties of polycrystalline $La_{1-x}Sr_xCoO_3$. *Journal of Solid State Chemistry*, 181, 3145–3150.

Kamihara, Y., Hiramatsu, H., Hirano, M., Kawamura, R., Yanagi, H., Kamiya, T., Hosono, H. (2006). Iron-based layered superconductor: LaOFeP. *Journal of the American Chemical Society*, 128, 10012–10013.

Kamihara, Y., Watanabe, T., Hirano, M., Hosono, H. (2008). Iron-based layered superconductor $La[O_{1-x}F_x]FeAs$ ($x = 0.05– 0.12$) with $T_c = 26$ K. *Journal of the American Chemical Society*, 130, 3296–3297.

Karppinen, M., Matvejeff, M., Salomäki, K., Yamauchi, H. (2002). Oxygen content analysis of functional perovskite-derived cobalt oxides. *Journal of Materials Chemistry*, 12, 1761–1764.

Kawasaki, S., Nakano, A., Taniguchi, H., Cho, H.J., Ohta, H., Nakamura, F., Terasaki, I. (2021). Thermal diffusivity of the Mott insulator Ca_2RuO_4 in a non-equilibrium steady state. *Journal of the Physical Society of Japan*, 90, 063601.

Kittel, C. (1996). *Introduction to Solid State Physics*. Wiley, New York.

Klein, L., Dodge, J., Ahn, C., Snyder, G., Geballe, T., Beasley, M., Kapitulnik, A. (1996). Anomalous spin scattering effects in the badly metallic itinerant ferromagnet $SrRuO_3$. *Physical Review Letters*, 77, 2774–2777.

Klein, Y., Hébert, S., Maignan, A., Kolesnik, S., Maxwell, T., Dabrowski, B. (2006a). Insensitivity of the band structure of substituted $SrRuO_3$ as probed by Seebeck coefficient measurements. *Physical Review B*, 73, 052412.

Klein, Y., Hébert, S., Pelloquin, D., Hardy, V., Maignan, A. (2006b). Magnetoresistance and magnetothermopower in the rhodium misfit oxide $[Bi_{1.95}Ba_{1.95}Rh_{0.1}O_4][RhO_2]_{1.8}$. *Physical Review B*, 73, 165121.

Knížek, K., Hejtmánek, J., Maryško, M., Jirák, Z., Buršík, J. (2012). Stabilization of the high-spin state of Co^{3+} in $LaCo_{1-x}Rh_xO_3$. *Physical Review B*, 85, 134401.

Kobayashi, W. and Terasaki, I. (2006). Transport properties of the thermoelectric layered cobalt oxide Pb–Sr–Co–O single crystals. *Applied Physics Letters*, 89, 072109.

Kobayashi, W., Ishiwata, S., Terasaki, I., Takano, M., Grigoraviciute, I., Yamauchi, H., Karppinen, M. (2005). Room-temperature ferromagnetism in $Sr_{1-x}Y_x\ CoO_{3-\delta}$ ($0.2 \leqslant x \leqslant 0.25$). *Physical Review B*, 72, 104408.

Kobayashi, W., Hébert, S., Pelloquin, D., Pérez, O., Maignan, A. (2007). Enhanced thermoelectric properties in a layered rhodium oxide with a trigonal symmetry. *Physical Review B*, 76, 245102.

Koshibae, W. and Maekawa, S. (2001). Effects of spin and orbital degeneracy on the thermopower of strongly correlated systems. *Physical Review Letters*, 87, 236603.

Koshibae, W. and Maekawa, S. (2003). Electronic state of a CoO_2 layer with hexagonal structure: A Kagomé lattice structure in a triangular lattice. *Physical Review Letters*, 91, 2570031–2570034.

Koshibae, W., Tsutsui, K., Maekawa, S. (2000). Thermopower in cobalt oxides. *Physical Review B*, 62, 6869–6872.

Koumoto, K., Terasaki, I., Funahashi, R. (2006). Complex oxide materials for potential thermoelectric applications. *MRS Bulletin*, 31, 206–210.

Koumoto, K., Funahashi, R., Guilmeau, E., Miyazaki, Y., Weidenkaff, A., Wang, Y., Wan, C. (2013). Thermoelectric ceramics for energy harvesting. *Journal of the American Ceramic Society*, 96, 1–23.

Kumagai, K., Suzuki, T., Taguchi, Y., Okada, Y., Fujishima, Y., Tokura, Y. (1993). Metal-insulator transition in $La_{1-x}Sr_xTiO_3$ and $Y_{1-x}Ca_xTiO_3$ investigated by specific-heat measurements. *Physical Review B*, 48, 7636–7642.

Kuroki, K. and Arita, R. (2007). "Pudding mold" band drives large thermopower in Na_xCoO_2. *Journal of the Physical Society of Japan*, 76, 083707–083707.

Lee, M., Viciu, L., Li, L., Wang, Y., Foo, M., Watauchi, S., Pascal Jr., R., Cava, R., Ong, N. (2006a). Large enhancement of the thermopower in Na_xCoO_2 at high Na doping. *Nature Materials*, 5, 537–540.

Lee, P.A., Nagaosa, N., Wen, X.-G. (2006b). Doping a Mott insulator: Physics of high-temperature superconductivity. *Reviews of Modern Physics*, 78, 17–85.

Liu, Y., Lin, Y.H., Zhang, B.P., Zhu, H.M., Nan, C.W., Lan, J., Li, J.F. (2009). High-temperature thermoelectric properties in the $La_{2-x}R_xCuO_4$ (R: Pr, Y, Nb) ceramics. *Journal of the American Ceramic Society*, 92, 934–937.

Macklin, W. and Moseley, P. (1990). On the use of oxides for thermoelectric refrigeration. *Materials Science and Engineering: B*, 7, 111–117.

Maignan, A., Hébert, S., Pi, L., Pelloquin, D., Martin, C., Michel, C., Hervieu, M., Raveau, B. (2002a). Perovskite manganites and layered cobaltites: Potential materials for thermoelectric applications. *Crystal Engineering*, 5, 365–382.

Maignan, A., Wang, L., Hébert, S., Pelloquin, D., Raveau, B. (2002b). Large thermopower in metallic misfit cobaltites. *Chemistry of Materials*, 14, 1231–1235.

Maignan, A., Eyert, V., Martin, C., Kremer, S., Frésard, R., Pelloquin, D. (2009). Electronic structure and thermoelectric properties of $CuRh_{1-x}Mg_xO_2$. *Physical Review B*, 80, 115103.

Masset, A., Michel, C., Maignan, A., Hervieu, M., Toulemonde, O., Studer, F., Raveau, B., Hejtmanek, J. (2000). Misfit-layered cobaltite with an anisotropic giant magnetoresistance: $Ca_3Co_4O_9$. *Physical Review B*, 62, 166–175.

Matsunaga, T., Kida, T., Kimura, S., Hagiwara, M., Yoshida, S., Terasaki, I., Kindo, K. (2010). High field magnetism of $Sr_{0.78}Y_{0.22}CoO_{3-\delta}$ under high pressure. *Journal of Low Temperature Physics*, 159, 7–10.

Merino, J. and McKenzie, R.H. (2000). Transport properties of strongly correlated metals: A dynamical mean-field approach. *Physical Review B*, 61, 7996–8008.

Miyazaki, Y., Kudo, K., Akoshima, M., Ono, Y., Koike, Y., Kajitani, T. (2000). Low-temperature thermoelectric properties of the composite crystal [$Ca_2CoO_{3.34}$]$_{0.614}$[CoO_2]. *Japanese Journal of Applied Physics*, 39, L531–L533.

Nakamura, F., Sakaki, M., Yamanaka, Y., Tamaru, S., Suzuki, T., Maeno, Y. (2013). Electric-field-induced metal maintained by current of the Mott insulator Ca_2RuO_4. *Scientific Reports*, 3, 2536.

Nakao, H., Murata, T., Bizen, D., Murakami, Y., Ohoyama, K., Yamada, K., Ishiwata, S., Kobayashi, W., Terasaki, I. (2011). Orbital ordering of intermediate-spin state of Co^{3+} in $Sr_3YCo_4O_{10.5}$. *Journal of the Physical Society of Japan*, 80, 023711.

Nakatsuji, S., Ikeda, S.-I., Maeno, Y. (1997). Ca_2RuO_4: New Mott insulators of layered ruthenate. *Journal of the Physical Society of Japan*, 66, 1868–1871.

Nishina, Y., Okazaki, R., Yasui, Y., Nakamura, F., Terasaki, I. (2017). Anomalous thermoelectric response in an orbital-ordered oxide near and far from equilibrium. *Journal of the Physical Society of Japan*, 86, 093707.

Ohta, S., Nomura, T., Ohta, H., Koumoto, K. (2005). High-temperature carrier transport and thermoelectric properties of heavily La-or Nb-doped $SrTiO_3$ single crystals. *Journal of Applied Physics*, 97, 034106.

Ohta, H., Kim, S., Mune, Y., Mizoguchi, T., Nomura, K., Ohta, S., Nomura, T., Nakanishi, Y., Ikuhara, Y., Hirano, M. (2007). Giant thermoelectric Seebeck coefficient of a two-dimensional electron gas in $SrTiO_3$. *Nature Materials*, 6, 129–134.

Ohtaki, M., Tsubota, T., Eguchi, K., Arai, H. (1996). High-temperature thermoelectric properties of ($Zn_{1-x}Al_x$)O. *Journal of Applied Physics*, 79, 1816–1818.

Okada, S. and Terasaki, I. (2005). Physical properties of Bi-based rhodium oxides with RhO_2 hexagonal layers. *Japanese Journal of Applied Physics*, 44, 1834–1837.

Okada, S., Terasaki, I., Okabe, H., Matoba, M. (2005). Transport properties and electronic states in the layered thermoelectric rhodate ($Bi_{1-x}Pb_x$)$_{1.8}Ba_2Rh_{1.9}O_y$. *Journal of the Physical Society of Japan*, 74, 1525–1528.

Okamoto, Y., Nohara, M., Takagi, H., Sakai, F. (2006). Correlated metallic phase in a doped band insulator $Sr_{1-x}Rh_2O_4$. *Journal of the Physical Society of Japan*, 75, 023704.

Okazaki, R., Nishina, Y., Yasui, Y., Shibasaki, S., Terasaki, I. (2011). Optical study of the electronic structure and correlation effects in $K_{0.49}RhO_2$. *Physical Review B*, 84, 075110.

Okazaki, R., Nishina, Y., Yasui, Y., Nakamura, F., Suzuki, T., Terasaki, I. (2013). Current-induced gap suppression in the Mott insulator Ca_2RuO_4. *Journal of the Physical Society of Japan*, 82, 103702.

Okazaki, R., Ikemoto, Y., Moriwaki, T., Nakamura, F., Suzuki, T., Yasui, Y., Terasaki, I. (2014). Disorder effect for an orbital order in Ca_2RuO_4 revealed by infrared imaging spectroscopy. *Journal of the Physical Society of Japan*, 83, 084701.

Okazaki, R., Kobayashi, K., Kumai, R., Nakao, H., Murakami, Y., Nakamura, F., Taniguchi, H., Terasaki, I. (2020). Current-induced giant lattice deformation in the Mott insulator Ca_2RuO_4. *Journal of the Physical Society of Japan*, 89, 044710.

Okuda, T., Nakanishi, K., Miyasaka, S., Tokura, Y. (2001). Large thermoelectric response of metallic perovskites: $Sr_{1-x}La_xTiO_3$ ($0<x<0.1$). *Physical Review B*, 63, 113104.

Palazzi, M., Carcaly, C., Flahaut, J. (1980). Un nouveau conducteur ionique (LaO)AgS. *Journal of Solid State Chemistry*, 35, 150–155.

Pálsson, G. and Kotliar, G. (1998). Thermoelectric response near the density driven Mott transition. *Physical Review Letters*, 80, 4775–4778.

Pelloquin, D., Maignan, A., Hébert, S., Martin, C., Hervieu, M., Michel, C., Wang, L., Raveau, B. (2002). New misfit cobaltites $[Pb_{0.7}A_{0.4}Sr_{1.9}O_3][CoO_2]_{1.8}$ (A = Hg, Co) with large thermopower. *Chemistry of Materials*, 14, 3100–3105.

Peterson, M.R. and Shastry, B.S. (2010). Kelvin formula for thermopower. *Physical Review B*, 82, 195105.

Prasad, R. and Bhame, S.D. (2020). Review on texturization effects in thermoelectric oxides. *Materials for Renewable and Sustainable Energy*, 9, 3.

Ramadass, N. (1978). ABO_3-type oxides – Their structure and properties – A bird's eye view. *Materials Science and Engineering*, 36, 231–239.

Ren, G.-K., Wang, S., Zhou, Z., Li, X., Yang, J., Zhang, W., Lin, Y.-H., Yang, J., Nan, C.-W. (2019). Complex electronic structure and compositing effect in high performance thermoelectric BiCuSeO. *Nature Communications*, 10, 2814.

Satake, A., Tanaka, H., Ohkawa, T., Fujii, T., Terasaki, I. (2004). Thermal conductivity of the thermoelectric layered cobalt oxides measured by the Harman method. *Journal of Applied Physics*, 96, 931–933.

Shibasaki, S., Kobayashi, W., Terasaki, I. (2006). Transport properties of the delafossite Rh oxide $Cu_{1-x}Ag_xRh_{1-y}Mg_yO_2$: Effect of Mg substitution on the resistivity and Hall coefficient. *Physical Review B*, 74, 235110.

Shibasaki, S., Nakano, T., Terasaki, I., Yubuta, K., Kajitani, T. (2010). Transport properties of the layered Rh oxide $K_{0.49}RhO_2$. *Journal of Physics. Condensed Matter*, 22, 115603.

Shibasaki, S., Terasaki, I., Nishibori, E., Sawa, H., Lybeck, J., Yamauchi, H., Karppinen, M. (2011). Magnetic and transport properties of the spin-state disordered oxide $La_{0.8}Sr_{0.2}Co_{1-x}Rh_xO_{3-\delta}$. *Physical Review B*, 83, 094405.

Shikano, M. and Funahashi, R. (2003). Electrical and thermal properties of single-crystalline $(Ca_2CoO_3)_{0.7}CoO_2$ with a $Ca_3Co_4O_9$ structure. *Applied Physics Letters*, 82, 1851–1853.

Singh, D.J. (2000). Electronic structure of $NaCo_2O_4$. *Physical Review B*, 61, 13397–13402.

Singh, D.J. (2003). Quantum critical behavior and possible triplet superconductivity in electron-doped CoO_2 sheets. *Physical Review B*, 68, 020503.

Slack, G.A. (1995). *CRC Handbook of Thermoelectrics*. CRC Press, Boca Raton.

Sudayama, T., Nakao, H., Yamasaki, Y., Murakami, Y., Asai, S., Okazaki, R., Yasui, Y., Terasaki, I. (2017). Spin state of Co^{3+} in $LaCo_{1-x}Rh_xO_3$ studied using X-ray absorption spectroscopy. *Journal of the Physical Society of Japan*, 86, 094701.

Sugano, S., Tanabe, Y., Kamimura, H. (1970). *Multiplets of Transition-Metal Ions in Crystals*. Academic Press, New York.

Tada, M., Yoshiya, M., Yasuda, H. (2010). Effect of ionic radius and resultant two-dimensionality of phonons on thermal conductivity in M_xCoO_2 (M = Li, Na, K) by perturbed molecular dynamics. *Journal of Electronic Materials*, 39, 1439–1445.

Takabatake, T., Suekuni, K., Nakayama, T., Kaneshita, E. (2014). Phonon-glass electron-crystal thermoelectric clathrates: Experiments and theory. *Reviews of Modern Physics*, 86, 669–716.

Takahashi, H., Ishiwata, S., Okazaki, R., Yasui, Y., Terasaki, I. (2018). Enhanced thermopower via spin-state modification. *Physical Review B*, 98, 024405.

Takashima, Y., Zhang, Y.-Q., Wei, J., Feng, B., Ikuhara, Y., Cho, H.J., Ohta, H. (2021). Layered cobalt oxide epitaxial films exhibiting thermoelectric $ZT = 0.11$ at room temperature. *Journal of Materials Chemistry A*, 9, 274–280.

Takeda, J., Nishikawa, T., Sato, M. (1994). Transport studies of $La_{1.92}Sr_{0.08}Cu_{1-x}M_xO_4$ (M=Ni and Zn) and $Nd_{2-y}Ce_yCuO_4$ up to about 900 K. *Physica C: Superconductivity*, 231, 293–299.

Tanabe, K., Taniguchi, H., Nakamura, F., Terasaki, I. (2017). Giant inductance in non-ohmic conductor. *Applied Physics Express*, 10, 081801.

Terasaki, I. (2001). Cobalt oxides and Kondo semiconductors: A pseudogap system as a thermoelectric material. *Materials Transactions*, 42, 951–955.

Terasaki, I. (2003). Transport properties and electronic states of the thermoelectric oxide $NaCo_2O_4$. *Physica B: Condensed Matter*, 328, 63–67.

Terasaki, I. (2011). High-temperature oxide thermoelectrics. *Journal of Applied Physics*, 110, 053705.

Terasaki, I., Sasago, Y., Uchinokura, K. (1997). Large thermoelectric power in $NaCo_2O_4$ single crystals. *Physical Review B*, 56, R12685–R12687.

Terasaki, I., Shibasaki, S., Yoshida, S., Kobayashi, W. (2010). Spin state control of the perovskite Rh/Co oxides. *Materials*, 3, 786–799.

Terasaki, I., Asai, S., Taniguchi, H., Okazaki, R., Yasui, Y., Ikemoto, Y., Moriwaki, T. (2017). Optical evidence for the spin-state disorder in $LaCo_{1-x}Rh_xO_3$. *Journal of Physics: Condensed Matter*, 29, 235802.

Thiel, P., Populoh, S., Yoon, S., Saucke, G., Rubenis, K., Weidenkaff, A. (2015). Charge-carrier hopping in highly conductive $CaMn_{1-x}M_xO_{3-\delta}$ thermoelectrics. *The Journal of Physical Chemistry C*, 119, 21860–21867.

Tokura, Y., Taguchi, Y., Okada, Y., Fujishima, Y., Arima, T., Kumagai, K., Iye, Y. (1993). Filling dependence of electronic properties on the verge of metal–Mott-insulator transition in $Sr_{1-x}La_xTiO_3$. *Physical Review Letters*, 70, 2126–2129.

Usui, H., Arita, R., Kuroki, K. (2009). First-principles study on the origin of large thermopower in hole-doped $LaRhO_3$ and $CuRhO_2$. *Journal of Physics: Condensed Matter*, 21, 064223.

Wang, Y., Rogado, N.S., Cava, R.J., Ong, N.P. (2003). Spin entropy as the likely source of enhanced thermopower in $Na_xCo_2O_4$. *Nature*, 423, 425–428.

Yamamoto, T.D., Taniguchi, H., Yasui, Y., Iguchi, S., Sasaki, T., Terasaki, I. (2017). Magneto-thermopower in the weak ferromagnetic oxide $CaRu_{0.8}Sc_{0.2}O_3$: An experimental test for the kelvin formula in a magnetic material. *Journal of the Physical Society of Japan*, 86, 104707.

Yamauchi, H., Karvonen, L., Egashira, T., Tanaka, Y., Karppinen, M. (2011). Ca-for-Sr substitution in the thermoelectric $[(Sr,Ca)_2(O,OH)_2]_q[CoO_2]$ misfit-layered cobalt-oxide system. *Journal of Solid State Chemistry*, 184, 64–69.

Zaanen, J., Sawatzky, G., Allen, J. (1985). Band gaps and electronic structure of transition-metal compounds. *Physical Review Letters*, 55, 418–421.

Zhao, L.D., Berardan, D., Pei, Y., Byl, C., Pinsard-Gaudart, L., Dragoe, N. (2010). $Bi_{1-x}Sr_xCuSeO$ oxyselenides as promising thermoelectric materials. *Applied Physics Letters*, 97, 092118.

Zhao, L.-D., He, J., Berardan, D., Lin, Y., Li, J.-F., Nan, C.-W., Dragoe, N. (2014). BiCuSeO oxyselenides: New promising thermoelectric materials. *Energy & Environmental Science*, 7, 2900–2924.

6
Nanocarbon Materials as Thermoelectric Generators

Tsuyohiko FUJIGAYA[1,2,3,4] and Yoshiyuki NONOGUCHI[5]
[1] Department of Applied Chemistry, Graduate School of Engineering,
Kyushu University, Japan
[2] The World Premier International Research Center Initiative, International Institute
of Carbon Neutral Energy Research, Kyushu University, Japan
[3] Department of Chemical Engineering, Graduate School of Engineering,
Kyushu University, Japan
[4] Center for Molecular Systems, Kyushu University, Japan
[5] Faculty of Materials Science and Engineering,
Kyoto Institute of Technology, Japan

6.1. Introduction

In the upcoming society, various sensing technologies are installed in our living space, and they are connected by wireless communication technologies, referred to as Internet of Things (IoT). The physical data carried by the communication network will be processed by artificial intelligence (AI) in cyberspace to add values to daily lives. Due to the huge number of sensors required, the exchange of batteries in, and wiring connection to, every sensor are not realistic, resulting in the expectation of energy harvesting technology. This chapter starts with the consideration of

---

For a color version of all the figures in this chapter, see www.iste.co.uk/akinaga/thermoelectric1.zip.

Thermoelectric Micro/Nano Generators 1,
coordinated by Hiroyuki AKINAGA, Atsuko KOSUGA,
Takao MORI and Gustavo ARDILA.
© ISTE Ltd 2023.

requirements for wireless power supply, using ambient waste heat. Thermoelectric (TE) conversion is a long-known technology for the interconversion of heat (temperature difference) and electrical energy. The power generation efficiency of TE power generation is the product of the Carnot efficiency and the dimensionless figure of merit, zT ($z = \sigma S^2 / \kappa$, σ: electrical conductivity (S m^{-1}), S: Seebeck coefficient (V K^{-1}), κ: thermal conductivity (W m^{-1}K^{-1}), T: absolute temperature (K)), where the higher the zT, the higher the power generation efficiency. The higher the zT, the higher the power generation efficiency. When a relatively large temperature difference of several hundred degrees Celsius is assumed (e.g. in furnaces and vehicles), good Carnot efficiency can be obtained, where thermoelectrics can be considered as a method to obtain renewable energy. This technology can potentially generate enough electric power to independently drive small electronic devices such as sensors and communications. There are many cases and infrastructures in our daily lives where we would like to set countless sensors in places where it is difficult to wire them to a public power supply or even to install solar cells. In this sense, energy harvesters including TE power generators are considered to be a powerful tool for realizing IoT society in near future.

Many high-efficiency TE inorganic materials and devices are available in the high-temperature range. On the other hand, the target of this chapter is to obtain microwatt-level power from temperature differences of a few to several tens of degrees Celsius in the environment, where thermal impedance matching between heat sources and TE generators is more and more important. TE generators are, in this context, required to closely attach to heat sources with a non-flat surface including pipes and human bodies. To address this issue, flexible TE modules are proposed in terms of two major directions (Bahk et al. 2015; Nan et al. 2018). Firstly, flexible substrates with movable electrodes are used for developing bendable TE modules. Another way for the same purpose is the use of flexible semiconductors and conductors including conducting polymers and carbon nanotubes (CNTs). In this chapter, we would like to introduce the recent progress in thermoelectric generators (TEGs) based on CNTs along with their basic studies on transport properties and carrier doping.

6.2. Carbon nanotubes

The driving temperature of TE materials is roughly correlated with the band gap, and it is known that bismuth telluride shows high TE efficiency around room temperature. Since the 2010s, many researchers have started to recognize that there are potentially useful TE materials in the material groups which are qualitatively different from the existing semiconductor and ceramics materials. The candidates include organic and nanocarbon materials such as conducting polymers and CNTs, where robust conjugated networks provide mechanically flexible, macroscopic architectures with designed electronic properties. Particularly, single-walled carbon nanotubes (SWCNTs) are light and structurally flexible even when formed into a relatively thick network film. From the viewpoint of electronic properties, they are one-dimensional semiconductors or metals made of carbon, and they transport electrons and heat efficiently. In the 1990s, it was pointed out that various physical properties such as the Seebeck effect can be controlled by nanostructuring (Hicks and Dresselhaus 1993). Due to the significant chemical and structural stability, in this context, SWCNTs are an intrinsically robust candidate for the investigation of low-dimensional thermoelectric transport, along with straightforward electronic and energy applications. Most of the commercially available CNTs are mixtures of semiconducting and metallic types depending on the helicity of graphite sheets. As far as the Seebeck coefficient is concerned, the mixture films are not much different from alloy metals such as constantan (30–50 $\mu V\ K^{-1}$). They can be controlled to be either p-type or n-type using rational chemical treatments, and then TE modules with a so-called π-type structure can be constructed to generate a reasonable amount of power.

6.3. Transport to materials studies

CNTs are produced by different suppliers with completely different properties. The current TE figure of merit (zT) is assumed to be in the range of 10^{-3} to 10^{-4} although there is a large variation among suppliers. The first observation of Seebeck coefficient (thermoelectric power, thermopower) in CNT mats was reported in 1998, by Zettl et al. (see Figure 6.1(a)) (Hone et al. 1998). Due to higher conductivity, the observed transport properties are considered to significantly reflect the nature of metallic CNT. The effects of environmental gases on the TE properties were then examined (see Figure 6.1(b)), where oxygen could induce p-type doping (Bradley et al. 2000). Afterwards, advanced fabrication and MEMS (micro electro mechanical systems) enabled the thermoelectric and thermal properties of individual carbon nanotubes (see Figure 6.1(c) and (d)), deepening the understanding from a physics point of view. Kim et al. (Small et al. 2003) revealed that TE power and transconductance in individual SWCNTs were significantly modulated by gate electric field and observed large TE power as high as ca. 260 µV K^{-1} at room temperature, derived from the Schottky barriers at SWCNT–metal junctions.

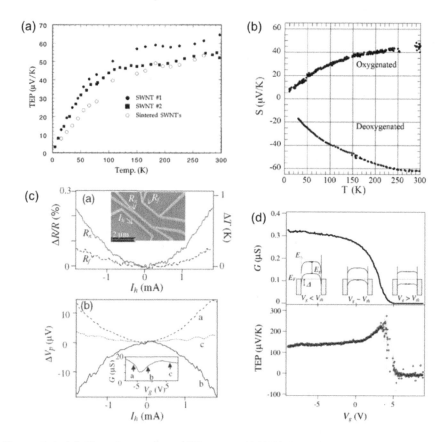

Figure 6.1. *(a) First observation of TE power of SWCNT samples. Reproduced with permission (Hone et al. 1998). Copyright 1998, American Physical Society. The temperature dependence of thermopower for a single sample in its "oxygenated" and "deoxygenated" states. Reproduced with permission (Bradley et al. 2000). Copyright 2000, American Physical Society. (c) The inset shows a scanning electron micrograph of a typical SWCNT device. The above graph shows the normalized change in Rn and Rf as a function of the heater current I_h at T = 300 K. Changes in resistance are mapped to the change in temperature difference ΔT across the SWCNT. (Bottom) TE voltage measured as a function of I_h at T = 300 K. Conductance (upper panel) and TEP (lower panel) versus gate voltage of a semiconducting SWCNT device at T = 300 K. The conductance plot exhibits typical p-type SWCNT FET behavior. The inset shows the band alignment of the SB model in different regions of the gate voltage. Reproduced with permission (Small et al. 2003). Copyright 2003, American Physical Society*

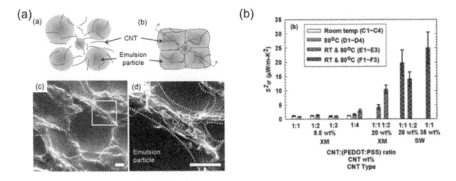

Figure 6.2. *(a) a. Schematic of CNTs suspended in an aqueous emulsion. Gray spheres and red lines represent emulsion particles and CNTs, respectively. b. Schematic of the emulsion-based composite after drying. The CNTs form a three-dimensional network along the surfaces of the spherical emulsion particles. SEMs of the cross-sections of 5 wt % CNT composites are shown in panels c and d after the composites were freeze-fractured (for instance, along the dotted line in panel b.). The high-magnification SEM shown in panel d is a portion of the sample in c indicated by a yellow solid square. It clearly shows that CNTs (indicated by arrows) are wrapped around the emulsion particles (indicated by yellow dotted lines) rather than homogeneously mixed. Denser CNTs were observed for higher-concentration CNT composites. The scale bars in the SEMs indicate 1 μm. Reproduced with permission (Yu et al. 2008). Copyright 2008, American Chemical Society. (b) Power factor ($S^2\sigma$) of the composites with 1:1, 1:2, 1:3 or 1:4 ratio between CNT and PEDOT:PSS composites with the fixed 9.8 wt % CNT concentration. Reproduced with permission (Kim et al. 2010). Copyright 2010, American Chemical Society*

In the early 2000s, due to low production ability and limited quality, SWCNTs were rarely recognized as TE materials for applications. The preparation of SWCNT films was empirically recognized to be difficult since SWCNTs at that time were too short to form spaghetti-like robust networks. In 2008, Yu et al. broke this limitation by preparing the composites of CNTs and emulsion particles (see Figure 6.2(a)) (Yu et al. 2008). Base polymers served as the support of CNTs, affording better percolation networks penetrating whole films. With a CNT concentration of 20 wt %, these composites exhibit an electrical conductivity of 4,800 S m^{-1}, a thermal conductivity of 0.34 W (m·K)$^{-1}$ and a TE figure of merit (ZT) greater than 0.006 at room temperature. This idea was spread as a breakthrough to the fabrication of various CNT-based composite materials around 2010. Chen et al. and Yu et al. have pioneered the CNT composites with conducting polymers such as polyaniline (Yao et al. 2010) and poly(3,4-ethylenedioxythiophene):poly(styrenesulfonate) (PEDOT:PSS) (Kim et al. 2010) yielding practical TE power factors as high as several tens μW $m^{-1} K^{-2}$ (see Figure 6.2(b)). State-of-the-art polymer composites made by layer-by-layer coating with high-quality dispersion was reported to exhibit a significant TE power factor: 1,825 $\mu W\ m^{-1} K^{-2}$ for polyaniline (PANi) (Cho et al. 2015), graphene, and double-walled CNTs, and 2,710 $\mu W\ m^{-1} K^{-2}$ for PANi/graphene-PEDOT:PSS/PANi/DWNT-PEDOT:PSS multilayers (Cho et al. 2016).

In the 2010s, significant progress in SWCNTs production and commercialization has resulted in their direct use along with a significant improvement in TE properties. Furthermore, the consequent scale-up of structure-sorted SWCNTs (e.g. semiconducting and metallic) has explored a new class of TE materials. Maniwa et al. discovered the giant Seebeck coefficient in semiconducting SWCNT film in the range of 160 $\mu V\ K^{-1}$ (see Figure 6.3(a)) (Nakai et al. 2014). An optimized power factor by acid doping reached 108 $\mu W\ m^{-1} K^{-2}$. In the work, they used the semiconducting and metallic SWCNTs sorted by density gradient ultracentrifugation. Another sorting method such as conducting polymer wrapping also enabled the systematic investigation of TE transport in highly pure (ca. 99%) semiconducting SWCNT thin films. Ferguson et al. tracked the doping progress by TE properties along with optical absorption, leading to an in-depth understanding in terms of SWCNT's one-dimensional electronic structures (Avery et al. 2016). Chemical doping systematically revealed the TE transport with an optimal power factor of approximately 340 $\mu W\ m^{-1} K^{-2}$ for the high-pressure carbon monoxide (HiPco) CNT networks. Additionally, they discovered significant suppression in thermal conductivity upon chemical doping, from ~16.5 $W\ m^{-1} K^{-1}$ down to 2–4.5 $W\ m^{-1} K^{-1}$, leading to synergistic enhancement in zT. They further improved the TE properties by using supramolecular sorting agents that can be washed by acid treatment (see Figure 6.4) (MacLeod et al. 2017). The thin film of semiconducting SWCNTs over 99% purity with a trace of surfactants was reported to show a huge power factor as high as ca. 705 $\mu W\ m^{-1} K^{-2}$ for the plasma-torch source, and zT~0.12 for the HiPco material. These values are still smaller than those of optimized bismuth telluride (zT~0.6). However, it is a breakthrough to find a high level of TE properties in SWCNTs, which can be made in the scale of several tens of tons worldwide.

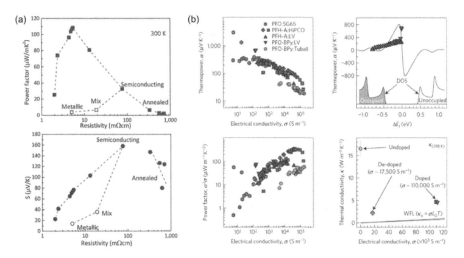

Figure 6.3. (a) Power factor and S as a function of volume resistivity at 300 K for acid-treated and vacuum-annealed semiconducting SWCNT films (filled symbols) and mixed and metallic SWCNT films. Reproduced with permission (Avery et al. 2016). Copyright 2014, IOP Publishing. (b) Experimentally measured TE properties as a function of electrical conductivity: thermopower and TE power factor for PFO:SG65 (7,5) (green hexagons), PFH-A:HiPCO (blue squares and diamonds), PFH-A:LV (red triangles), PFO-BPy:LV (inverted purple triangles) and PFO-BPy:Tuball (orange circles). Comparison of theoretical (solid line) and experimental (colored symbols) dependence of the thermopower on the position of the Fermi energy for LV s-SWCNTs (the theoretically predicted electronic DOS is shown for reference). Thermal conductivity of a PFO-BPy:LV thin film as a function of the electrical conductivity near 300 K. Reproduced with permission (Avery et al. 2016). Copyright 2016, Springer Nature Limited

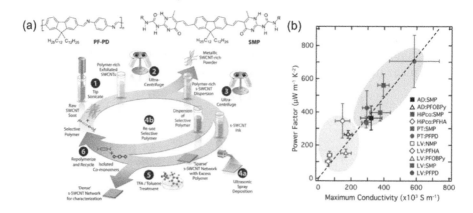

Figure 6.4. *(a) Generalized process for making high-performance s-SWCNT TE thin films with no residual sorting polymer. (b) Dependence of maximum TE power factor (optimally doped network) on maximum conductivity (fully doped network) for 11 unique s-SWCNT networks. The maximum conductivity and TE power factor for all networks using cleavable polymers (blue oval) exceed the values for networks containing residual wrapping polymer (orange oval). Reproduced with permission (MacLeod et al. 2017). Copyright 2017, The Royal Society of Chemistry*

6.4. Chemical doping

Among TE properties, Seebeck coefficient, electrical conductivity and thermal conductivity are functions of carrier density or Fermi level. It has also been shown that there is a trade-off between these three factors in improving zT. Therefore, controlling the carrier density, or carrier doping, is important in designing TE materials. In addition, TE conversion modules with high efficiency have a so-called π-type structure, in which p-type and n-type materials are sandwiched between electrodes. This structure is superior to the monopolar structure in terms of heat utilization efficiency. For physical properties research, carrier doping has been performed by electrochemical doping, where TE measurements were performed in situ (Yanagi et al. 2014). On the other hand, for practical use, we need a doping method that can be applied stably and reproducibly on a stand-alone basis. In the following sections, the authors will introduce several approaches to practical doping.

Doping in semiconductor engineering generally involves the introduction of ions of different valences. On the other hand, it is believed that nanocarbons can be implanted with carriers by substituting heteroatoms such as boron and nitrogen for carbon, but the mass production technology is still in the research stage and there are many problems in controlling the substitution sites. Therefore, carrier injection using oxidants and reducing agents has been investigated.

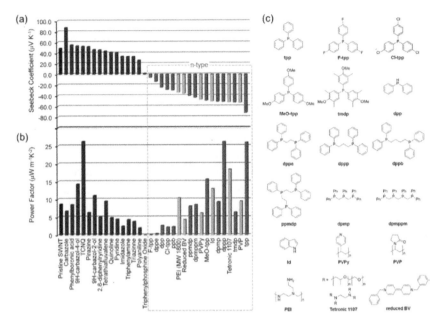

Figure 6.5. *Modulation of the Seebeck coefficient by adsorption doping. (a) Seebeck coefficients of SWCNT films doped with each molecule above, at 310 K. (b) Power factors of pristine and doped SWCNT films at 310 K. Abbreviations are referred to (c). Reproduced with permission (Nonoguchi et al. 2013). Copyright 2013, Springer Nature Limited*

The n-type doping of CNTs was initially confirmed by the addition of very intense reducing agents such as potassium metal, but potassium metal is extremely dangerous and unsuitable for realistic processes. Adsorption of poly(ethyleneimine) (PEI) has also been used to n-type SWCNTs, but the doping effect seems to vanish after a few days of exposure to air. This degradation is more remarkable for semiconducting SWCNTs having a shallow conduction band than the mixture of semiconducting and metallic SWCNTs (Shim et al. 2001). This is probably due to the low stability of radical anion-type bonding in the graphene framework produced by electron doping, and the reaction of unstable sp3-like carbons with unshared electrons with electron uptakers such as oxygen (O_2), carbon dioxide (CO_2) and water (H_2O) under atmospheric conditions.

In 2013, Nonoguchi et al. prepared the composite sheets of SWCNTs and the phosphine-based compounds represented in Figure 6.5 and discriminated the majority carriers by Seebeck coefficient (Nonoguchi et al. 2013). A positive value in the Seebeck coefficient indicates a p-type material, while a negative value indicates an n-type material. The undoped Seebeck coefficient was +49 µV K^{-1} at 310 K. After molecular doping, the Seebeck coefficient changed significantly from +90 µV K^{-1} to –80 µV K^{-1}. In particular, the electron-donor-doped CNTs in Figure 6.5(c) showed a negative Seebeck coefficient, indicating the realization of stable n-type CNTs. In addition to amine-based molecules, a number of phosphine (phosphorus) derivatives were found to give relatively stable n-type CNTs. A clear correlation was found between the work function of the film and the electron-donatable level (HOMO) of the dopant compound, as measured by atmospheric photoelectron yield spectroscopy, strongly suggesting that intermolecular charge transfer following adsorption is the origin of the doping.

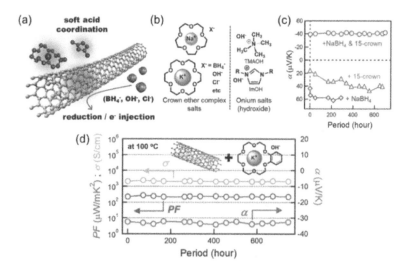

Figure 6.6. (a) Naked anion reduction to CNTs. (b) Crown ether complexes and onium salts inducing the naked anion effect. (c) Temporal changes of the Seebeck coefficient (α) of SWCNT films treated solely with $NaBH_4$ or 15-crown-5-ether, and with both $NaBH_4$ and 15-crown-5-ether over a month under ambient conditions. (d) Stability of TE properties of n-type CNT films with benzo-18-crown-6-ether complexes accelerated at 100°C in air. Reproduced with permission (Nonoguchi et al. 2018). Copyright 2016, John Wiley & Sons, Inc.

The phosphine-treated SWCNTs afforded 2.5-fold enhanced the TE power factor (ca. 25 μW m^{-1} K^{-2}) compared to that of the PEI-treated SWCNTs, and zT increased by about five times. The n-type CNTs treated with triphenylphosphine were stable in air for about a month, but were sensitive to heat and humidity. It is important to note that the phosphine/phosphonium redox gave more stable n-type CNTs than the amine/ammonium redox. This means that CNT anions were stabilized by larger cations such as phosphonium.

Considering the successful progress in the n-type SWCNTs' stability above, it is likely that the introduction of larger molecular cations is beneficial for achieving more thermodynamic stability. In this context, Nonoguchi et al. developed a supramolecular system that enables the electron doping of SWCNTs and the in situ formation of appropriate molecular counterions for n-type SWCNTs (see Figure 6.6(a) and (b)) (Yoshiyuki et al. 2016). Typically, they used an alcoholic solution of crown ether and potassium hydroxide at around 10~100 mM. Crown ether is a typical supramolecular compound and is known to accommodate metal ions in its own pocket in high yields. After immersion and drying of CNT films in this solution, they found that stable n-type films were prepared under atmospheric conditions. As a control, we confirmed that stable n-type CNT films could not be obtained with crown ether and potassium hydroxide solutions alone. The present method produced relatively more stable n-type films than the conventional method (see Figure 6.6(c)). Through the optimization of the nanotube structure and dopant concentration, they achieved the preparation of films which did not show significant degradation for one month under accelerated test conditions of about 100°C in air (see Figure 6.6(d)). The optimized TE power factor exceeded 220 μW m^{-1} K^{-2}, about 10 times larger than that reported in 2010–2013 due to the availability of high-quality nanotubes at that time and the dramatic improvement in film preparation technology including doping.

In this reaction solution, the hydroxide ion is destabilized by being separated from the potassium ion due to the shielding effect of the crown ether (the naked anion effect). When the destabilized hydroxide ions act on the carbon nanotubes, the reduction reaction proceeds reproducibly, and electrons are injected into the nanotubes. The fact that this is not a degradation reaction but a doping of carbon nanotubes is confirmed by the modulation of interband absorption and plasmon resonance, the dramatic increase in conductivity and the positive and negative Seebeck coefficient reversals.

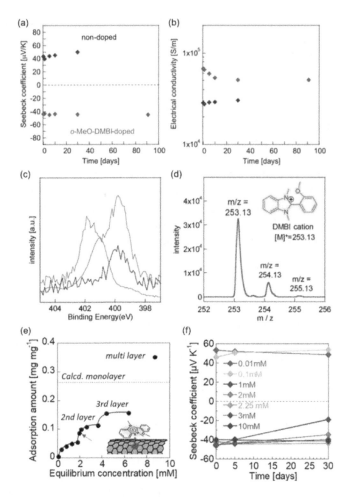

Figure 6.7. Time course of (a) Seebeck coefficient and (b) electrical conductivity of the non-doped (black) and o-MeO-DMBI-doped (red) SWCNT film measured at 48°C. (c) XPS N1s narrow scans of o-MeO-DMBI (blue line), the non-doped SWCNT film (black line) and the o-MeO-DMBI-doped SWCNT film (red line). (d) MALDI-TOF mass spectrum of the o-MeO-DMBI-doped SWCNT films without adding any matrix regent. Reproduced with permission (Nakashima et al. 2017). Copyright 2016, Elsevier B.V. (e) Adsorption isotherm of o-MeO–DMBI on SWCNT sheet measured at 25°C. The line was added as a guide. The red dotted line indicates the calculated adsorption amount based on the BET surface area of SWCNT sheet. The red arrow indicates the minimum concentration (2.25 mM) realizing air-stable n-doping. (f) Time course of the Seebeck coefficient of SWCNT sheets doped for 50 h with various concentrations of o-MeO–DMBI solutions. Reproduced with permission (Nakashima et al. 2019). Copyright 2019, American Chemical Society

In addition, the complexes of potassium ion and crown ether are adsorbed around the negatively charged carbon nanotubes to satisfy the charge neutrality rule. It is known that the reduction reaction of potassium/ potassium ions does not produce stable n-type carbon nanotubes, and therefore the contribution of the crown ether complex to the stabilization is strongly suggested. The charges on the CNTs can be delocalized to the two-dimensional plane of the graphitic framework, and their stabilization by planer crown complex cations is consistent with the hard/soft acid–base (HSAB) rule. The structure dependence of the cations was investigated by electrochemical doping, confirming that macrocyclic crown ethers were effective (Nonoguchi et al. 2018). It was also found that large π-conjugated organic cations other than crown ethers also gave stable n-type films (Nonoguchi et al. 2017a, 2017b), and the same concept was successfully applied for a p-type counterpart (Nakano et al. 2017).

Fujigaya et al. developed another way with organic reductants for the stabilization of n-type SWCNTs, expanding the design principle for n-type doping towards practical module fabrication (Nakashima et al. 2017). Particularly, they found that 2-(2-methoxyphenyl)-1,3-dimethyl-2,3-dihydro-1H-benzo[d]imidazole (o-MeO–DMBI) offers air-stable n-type SWCNT sheets. The wet-chemically prepared material showed n-type TE properties over 90 days (see Figure 6.7(a) and (b)). In the n-type doping reaction, o-MeO–DMBI cations were detected as a side product by X-ray photoelectron spectroscopy (XPS) and MALDI-TOF mass spectroscopy (see Figure 6.7(c) and (d)), suggesting the cations could compensate the negative charges injected in SWCNTs. Furthermore, they found unique stability dependence on dopant concentration, where the number of molecular dopant layers is crucial for obtaining effectively stable n-type SWCNT films (Nakashima et al. 2019). Adsorption isotherm measurements were applied to check the molecular layers, revealing that the entire coverage of the SWCNT surface by o-MeO–DMBI cation is the key requirement to realize the air stability. In a similar manner, various organic reductants including not only hydrogen atom or hydride transfer agents but also in situ generated radicals (Tanaka et al. 2021) are expected to serve for clean n-type doping of SWCNTs. Due to flexibility in the dopant design with organic reductants, the dopant introduction methods are not limited in the wet-chemical way such as casting and soaking but make it possible to use dry processes including vacuum deposition associated with the more precise control of doping positions and areas. Their development will pave the way for constructing design principles for arbitrary TE devices based on SWCNT power generation layers.

6.5. Thermoelectric generators using CNT

Thermoelectric generators (TEGs) using TE materials are attractive because they do not require moving parts and can generate electricity silently by using a temperature gradient. A TEG module is the π-shaped assembly of alternating p-type and n-type TE materials, which are connected electrically in series and thermally in parallel (Nan et al. 2018). An advantage of using CNTs for TEGs is their bendability, which imparts flexibility or even stretchability to the device. Therefore, a TEG module that can cover the surface of heat sources and take complete advantage of ubiquitous waste heat, such as body heat, has been the target of the CNT-based TEG. Recent studies revealed that the zT of CNT-based materials is still lower than 0.3 (Blackburn et al. 2018); thus, a large integration of the p- and n-type CNT legs is vital to achieve considerable power generation. To increase power generation, many TEG structures have been developed based on CNT and CNT/polymer composites. These structures were designed based on those already developed for either inorganic or organic TEGs (Nandihalli et al. 2020). However, CNTs possess unique one-dimensional (1D) structures and CNT-based materials often exhibit strong electrical and thermal conduction anisotropy based on their orientation. Thus, TEG design and preparation methods always require full consideration of their anisotropy. The electrical and thermal conductivities of CNT sheets in the in-plane direction are more than 100 times higher than those in the out-of-plane direction (Wei et al. 2014). Although the structures of TEGs based on CNTs are similar to those of inorganic bulk materials or organic materials, their preparation methods are largely different. For example, in conventional inorganic bulk materials, TEG modules are fabricated by dicing the ingot, assembling the diced pieces on the substrate, and connecting it with an electrode. In comparison, for the CNT-based TEG module, CNT dispersion and CNT sheet preparation via filtration, doping and assembly into a p-n assembly were carried out.

To produce sufficient power, p- and n-type TE materials need to be assembled in the TEG module to fully use the temperature gradient (ΔT) available from the environment. Typically, TEG modules composed of bulk materials such as Bi_2Te_3 are arranged such that the temperature difference between the two ends of the TE materials is perpendicular to the surface of the module. Conversely, because the CNT or CNT/polymer composites are typically thin films, the availability of the temperature difference in the out-of-plane direction is limited. Therefore, the three-dimensional (3D) structure of the TEG was investigated for CNTs. In the following sections, the developments of both 2D and 3D TEG structures are described.

6.6. TEG based on CNT sheet

Because CNT sheets are prepared easily by vacuum filtration of the CNT dispersion and can be cut with scissors, the assembly of multiple CNT sheets with arbitrary sizes has been tested. In 2012, Yu et al. demonstrated a simple π-shape 2D TEG composed of PEI-doped (n-type) and undoped (p-type) single-walled SWCNT sheets (Figure 6.8(a)) (Yu et al. 2012). The three p–n couples exhibited a power of 25 nW at a $\Delta T = 22$ K. In 2013, Nonoguchi et al. fabricated a bendable TEG module on a thin polyimide film using SWCNT sheets doped with triphenylphosphine and tetracyanoquinodimethane (TCNQ) as n- and p-type legs, respectively (Nonoguchi et al. 2013). The three p- and n-type legs were electrically connected in series using flexible copper tapes. The module generated a power of 110 nW at a ΔT of 20 K, which agreed with the values predicted by the Seebeck coefficients of both p- and n-type materials.

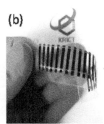

Figure 6.8. *(a) 2D π-shaped TEG fabricated from undoped SWCNT and PEI-doped SWCNT legs. Reproduced with permission (Yu et al. 2012). Copyright 2012, The Royal Society of Chemistry. (b) A flexible TEG on PET substrate; 10 p–n couples composed of F_4TCNQ-doped CNT and BV-doped CNT. Reproduced with permission (An et al. 2017). Copyright 2017, The Royal Society of Chemistry*

In 2017, An et al. demonstrated high power generation from a 2D π-shaped TEG consisting of p-type CNT legs doped with 2,3,5,6,-tetrafluoro-7,7,8,8-tetracyanoquinodimethane (F_4TCNQ) and n-type CNT legs doped with benzyl viologen (BV) dichloride (An et al. 2017). Freestanding p-type and n-type CNT webs exhibiting high power factors of 2,252 and 3,103 $\mu W \cdot m^{-1} \cdot K^{-2}$, respectively, were attached onto a flexible PET substrate and connected electrically in series using silver paste. A flexible TEG with 10 p–n couples generated 7.1 μW of power at a $\Delta T = 20$ K, corresponding to an output per weight of 28.3 $\mu W \cdot g^{-1}$ and output per occupying area of 2.0 $\mu W \cdot cm^{-2}$ (Figure 6.8(b)). When the area was normalized to the leg cross-sectional area, a high power density of 1,180 µW·cm^{-2} was obtained.

A promising strategy to improve power density is to extend the structure in the vertical direction, instead of the lateral direction. Hewitt et al. (2012) and Kim et al. (2014) demonstrated 3D TEGs based on a multilayer stack of CNT sheets that used an in-plane temperature gradient. Kim et al. stacked non-doped CNT sheets (p-type) and n-type CNT sheets doped with PEI, diethylenetriamine and NaBH$_4$ alternately in series with insulating films of polytetrafluoroethylene (PTFE) inserted between them (see Figure 6.9). The TEG modules containing 72 p–n couples of the CNT films produced a power output of 1.8 µW at a $\Delta T = 32$ K.

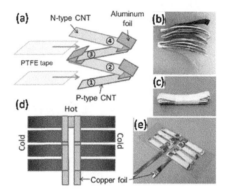

Figure 6.9. *(a) Multilayer stack of p- and n-type CNT films with insulating PTFE films. (b) Photo of one TEG unit consisting of nine p-type and nine n-type films. (c) Photo of the stack after being bound by a PTFE tape. (d) A device design consisting of eight TEG units and (e) Photo of a TEG module consisting of eight TEG units. Reproduced with permission (Kim et al. 2014). Copyright 2014, American Chemical Society*

In the above π-shaped configurations, the p- and n-type legs are connected with conductive pastes or metal electrodes. However, these conductive pastes act as electrical resistances, thus lowering the output power of TE devices. Mytafides et al. pointed out that the contact resistance of the metals/SWCNT interface is higher than that of highly conductive SWCNTs and demonstrated that the power output of the joint-free structure is higher than that of the Ag-interconnected TEG with the same architecture (Mytafides et al. 2021).

In 2017, Zhou et al. reported joint-free TEG devices wherein CNT films (96 mm × 10 mm) with a thickness of ~3 µm were patterned into p- and n-type regions to eliminate the metal joint (Zhou et al. 2017). The SWCNT sheet (p-type) transferred onto a PET substrate was alternately masked with thin PET double-side adhesive taps, and the unmasked regions were doped by drop-casting PEI to produce n-type legs. Then, a joint-free p–n patterned sheet was folded, and a compact multilayer p–n stuck structure with dimensions of 16 mm × 10 mm was fabricated (see Figure 6.10). Due to the relatively high power factor of these CNT legs (p-type: 1,840 µW·m^{-1}·K^{-2} and n-type: ~1,500 µW·m^{-1}·K^{-2}) as well as the joint-free structure, the modules consisting of three p–n couples produced a maximum power of 2.51 µW at a ΔT of 27.5 K, which corresponds to a power density of 0.26 µW·cm^{-2} (167 µW·cm^{-2} when normalized to the total area and the cross-sectional area).

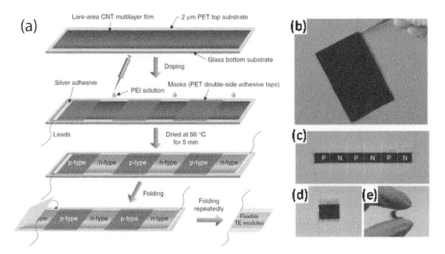

Figure 6.10. *(a) Schematic illustration of the fabrication process. (b) Photo of the original CNT film. (c) Photo of a CNT composed of three couples of continuous p–n couples. (d) Photo of a compact TEG module after folding. (e) Photo of a TEG module showing the flexibility. Reproduced with permission (Zhou et al. 2017). Copyright 2017, Nature Publishing Group*

In these 2D π-shaped modules, the temperature gradient must be applied in the in-plane direction. However, as most of the waste heat is emitted in the out-of-plane direction (e.g. body heat), such a π-shaped configuration limits its application. Thus, TEG modules that can harvest heat in the out-of-plane direction are required. Therefore, 3D TEGs with vertically aligned TE legs have been proposed. An et al. demonstrated a 3D vertical TEG by stacking 20 layers of 10 2D p–n couples to fully use the temperature gradient in the vertical direction (see Figure 6.11), and a maximum power output of 123 µW was achieved at a ΔT of 20 K, which means that the power output of the module normalized by the number of couples was 0.6 µW (An et al. 2017).

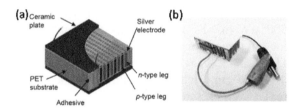

Figure 6.11. *(a) Schematic illustration and (b) photograph of the 3D TEG module consisting of 200 p–n couples. Reproduced with permission (An et al. 2017). Copyright 2017, The Royal Society of Chemistry*

Using the foldable feature of CNT sheets, Kim et al. also developed a 3D structure of the CNT TEG. The paper sheet with a length of 1.5 cm and a breadth of 24 cm (area 36 cm^2) having 38 p-type and 37 n-type CNT legs was folded using a double-sided adhesive tape (see Figure 6.12) (Kim et al. 2019). The folded legs with a length of 1.5 cm and projected area of 2.25 cm^2 exhibit maximum powers of 1.1, 4.4 and 10.3 µW at the vertical ΔT values of 10, 20 and 30 K, respectively. Based on the open-circuit voltages, they estimated that approximately 50% of the ΔT was used in the 3D vertical TEG. The maximum power densities of the device normalized by the projected area (2.25 cm^2) were calculated to be 0.5, 2.0 and 4.6 µW cm^{-2} at the vertical ΔT values of 10, 20 and 30 K, respectively, but those normalized by the area of the paper (36 cm^2) were 0.03, 0.12 and 0.29 µW cm^{-2}, respectively. These values are lower than those of the 2D CNT TEG reported by An et al. (2.0 µW·cm^{-2} at ΔT = 20 K) (An et al. 2017), which is likely due to lower power factors (p-type: 411 µW·m^{-1}·K^{-2}; n-type: 90.2 µW·m^{-1}·K^{-2}) of the TE sheet used compared to those of An et al. (p-type: 2,252 µW·m^{-1}·K^{-2}; n-type: 3,103 µW·m^{-1}·K^{-2}). Additionally, in the case of 2D TEG sheets, both the cold and hot sides were temperature controlled by an external temperature controller, and the temperature gradient could be fully used. The cold side of the 3D vertical TEG is usually uncontrolled (cooled by air atmosphere), and the temperature is often different from room temperature. Hence, difficulty of fair comparison in this area arises.

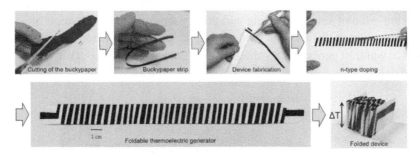

Figure 6.12. *Photographs representing the fabrication process of the foldable TEG. Reproduced with permission (Kim et al. 2019). Copyright 2019, American Chemical Society*

Lee et al. developed a 3D CNT TEG using porous CNT foams, where SWCNT slurries were poured into a polydimethylsiloxane (PDMS) mold and subjected to drying at a reduced pressure to form porous foams (Lee et al. 2019). SWCNT slurries were doped with $FeCl_3$ and BV for p- and n-type doping, respectively, and the doped p-and n-type legs were connected using silver paste. In this approach, thick films were easily prepared with a small amount of CNTs because of the large porous structure, which helps in using the applied vertical direction temperature gradients. As a result, a maximum power output of 1.5 µW at a ΔT of 13.9 K was obtained, corresponding to a high output power by weight of 82 $\mu W \cdot g^{-1}$.

Dörling et al. proposed a unique toroidal geometry to convert a 2D flat film into a 3D architecture. The structure was prepared by connecting the p- and n-type legs with silver paste, folding into a spiral with adjacent couples (electrically in series) and finally joining to form a torus (see Figure 6.13) (Dörling et al. 2016). Because of this 3D architecture, the temperature difference between the skin surface and air can be used. The fabrication of joint-free TEG devices requires patterning of the p- and n-type legs to fabricate alternative assemblies of p- and n-type areas onto CNT sheets. However, both drop-casting and brushing/printing techniques resulted in a low resolution of the patterning due to the capillarity effect of the fibrous CNT network structure (Sun et al. 2020). On the other hand, they used photo-induced switching from p-type to n-type for patterning. Upon UV irradiation, an initially p-type composite comprising P3HT and nitrogen-doped MWCNTs changed to n-type. This approach can avoid the diffusion issue of solution doping and greatly simplify the production of TEGs. When one side of this module with 15 double legs was attached to a glass filled with ice water, it generated a voltage of 5 mV and power of 10 μW.

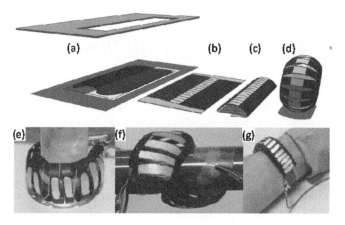

Figure 6.13. (a–c) Fabrication process of the TEG based on photo-induced patterning of the n-doped region. (d) Final structure of the 3D toroidal TEG. Possible application geometries in the form of (e) a single torus, (f) an extended spiral and (g) a wristband. Reproduced with permission (Dörling et al. 2016). Copyright 2016, Wiley-VCH

Yamaguchi et al. developed a novel approach for the n-doping of SWCNT sheets by thermal evaporation of o-MeO-DMBI iodide and used this approach for the patterning of SWCNT sheets into p-n couples (Yamaguchi et al. 2021). In this TEG module, a patterned 2D SWCNT sheet was designed to generate power in the in-plane direction by harvesting the temperature gradient in the out-of-plane direction. To realize this concept, a thermal conductor (Cu wire) was attached between the p- and n-type regions of the patterned SWCNT sheet, and the other area was passivated by a thermal insulator (see Figure 6.14). Four p–n couples of the 2D CNT TEG generated a power of 96.1 nW, corresponding to a power density of 60 nW cm^{-2}. In this concept, minimizing the pattern pitch can further increase the power density.

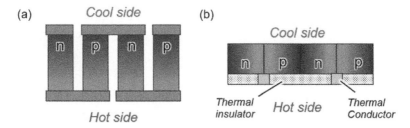

Figure 6.14. *(a) Typical 2D π-shaped TEG structure and (b) planner-type 2D TEG. Reproduced with permission (Yamaguchi et al. 2021). Copyright 2021, The Royal Society of Chemistry*

6.7. TEG fabrication based on CNT-based ink

An advantage of CNTs is their solution processability when dispersed in solvents. The solution can be used as an "ink" to fabricate TEG structures through a variety of methods such as screen printing, inkjet printing, dispenser printing, spray printing and spin coating (Nandihalli et al. 2020) (Hong et al. 2015). To prepare CNT ink with good processability, de-bundling of CNTs is highly important; thus, the addition of dispersants is necessary. For this reason, CNT inks often contain polymers or small surfactants as dispersants.

Suemori et al. reported a pioneering work that described the fabrication of TEGs on flexible film substrates using CNT/polystyrene composite ink dispersed in 1,2-dichlorobenzene (see Figure 6.15) (Suemori et al. 2013). They fabricated a TEG with CNT ink through a screen printing process. After the composite solution was dried, the printing mask was removed from the substrate and a gold electrode was attached on top by vacuum deposition. The TEG composed of 1,985 legs (1.5 mm × 0.8 mm × 0.15 mm) was operated in the out-of-plane direction, and a power output of approximately 5.5 µW cm^{-2} at a ΔT of 70 K was achieved.

In 2015, Hong et al. fabricated a 2D π-shaped TEG by spray printing through a shadow mask using a p-type CNT/P3HT composite ink on a polyimide substrate (Hong et al. 2015). In this TEG, 41 p-type lines with a width of 1 mm and a length of 15 mm were connected in series using a silver electrode. This device produced a power of 32.7 nW at a ΔT of 10 K.

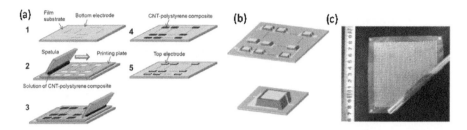

Figure 6.15. *(a) Schematic illustration of the fabrication of the TEG based on screen printing method using CNT ink. (b) Structure of the TEG fabricated in this study (upper panel) together with the close-up illustration shows structure of the individual device (lower panel). (c) Photograph of the TEG composed of 1,985 individual TE legs on a polymer film substrate. Reproduced with permission (Suemori et al. 2013). Copyright 2013, AIP Publishing LLC*

Although these works fabricated TEG only with p-type CNT legs, to fabricate an efficient TEG module, n-type legs need to be connected alternatively to minimize parasitic electrical resistance and to facilitate the heat transfer into the system; thus, a doping technique to convert p-type CNTs into n-type is necessary. When the proper dispersants are chosen, the dispersants work as a dopant to tune p-type into n-type; thus, post-doping after the printing process can be avoided. Such advantages enable various TEG structures composed of multiple p–n assemblies in a simple printing process. To produce high-quality printable TE inks with high performance, it is important to produce high-quality dispersions with the appropriate viscosity depending on the design, application and substrate.

In 2018, Park et al. designed a bracelet-type TEG structure wherein the CNT ink was printed onto a flexible polyurethane cable (see Figure 6.16) (Park et al. 2018). CNT inks were prepared using polyacrylic acid (PAA) and PEI as p- and n-dopants, respectively, in solvents such as diethylene glycol. By optimizing the concentration of the dopants and solvents, clogging of inks in the nozzle can be avoided while making them sufficiently viscous for printing on any curved surface. Because of the 3D structure, the device was operated in the out-of-plane direction of the heat source. The flexible TEG based on 60 couples of p- and n-doped CNT ink obtained a maximum power output of 1.95 μW at a ΔT of 30 K. The ease of installation of the bracelet-type TEG on heat sources with various shapes and its ability to harvest waste heat in the out-of-plane direction of the heat source has significant potential as a flexible/wearable power conversion device. Because of the ease of printing, the number of p–n pairs can be easily increased without any additional assembly process.

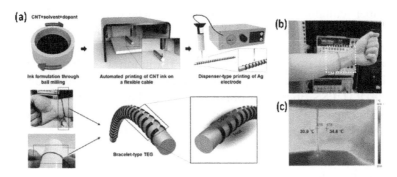

Figure 6.16. *(a) Illustration showing the fabrication of a bracelet-type TEG by printing of CNT inks onto a flexible cable. (b) Bracelet-type TEG harvesting waste heat on any curved surface. (c) Demonstration of the bracelet-type TEG wrapped around the wrist. Reproduced with permission (Park et al. 2018). Copyright 2018, The Royal Society of Chemistry*

Mytafides et al. used dodecylbenzene sulfonate (SDBS) and cetyltrimethylammonium bromide (CTAB) as dispersants, as well as the dopant of SWCNTs in water and prepared p- and n-type aqueous inks, respectively (Mytafides et al. 2021). Initially, the p-type CNT ink was printed on a flexible polyimide substrate using a blade coating technique to prepare p-type legs, and then, using the same technique, the n-type ink was printed to create a continuous electric path consisting of 116 p–n legs. It is also worth noting the remarkable power factors of 145 and 127 µW mK^{-2} achieved for the p-type and n-type films, respectively, at room temperature. In this configuration, highly conductive SWCNT networks are also employed as joints between the p–n regions, which realize an all-carbon TEG without metal deposition. As a result, the 2D π-shaped TEG module exhibited a remarkable power output of 342 µW at a ΔT = 150 K (T_H = 175°C) with an internal resistance of 806 Ω. The use at high temperatures (up to 200°C) is one of the advantages of CNTs with excellent thermal stability.

6.8. CNT yarn and their fabric

CNT or CNT/polymer composite fibers can be fabricated from CNT ink or CNT/polymer composite inks. These fibers are also candidates for the fabrication of TEG modules. Kim et al. reported a TEG fabricated from a CNT fiber composited with PEDOT:PSS, where the CNT/PEDOT:PSS fibers were synthesized by ball milling the dispersion of SWCNTs by PEDOT:PSS to make an ink paste and then spinning the paste in methanol for coagulation (Kim et al. 2018). The PEI solution was filtered through the CNT/PEDOT:PSS fibers to prepare n-type fibers. The p-type and n-type CNT/PEDOT:PSS fibers thus obtained exhibited power factors of 83.2 ± 6.4 and 113 ± 25 $\mu W \cdot m^{-1} \cdot K^{-2}$, respectively. As a result, 12 p–n couples of the TEG obtained an output power of 0.430 µW at a ΔT of 10 K.

Recently, electronic textiles (e-textiles) have attracted increasing attention for the development of smart textiles to enhance textile functionalities including sensing, energy harvesting and active heating and cooling. For this purpose, the integration of TEG into textiles has been investigated using various TE materials, including inorganic, organic and their hybrids. In particular, CNT-based TE materials are attractive because of their ease of fiber fabrication and textile compatibility.

In 2017, Ito et al. (2017) first demonstrated the integration of CNT threads into fabrics (see Figure 6.17(a)). To fabricate the CNT thread, surfactant-dispersed SWCNTs were mixed with polyethylene glycol to reinforce the thread, and the dispersion was injected into methanol to induce coagulation. One of the technical issues in integrating CNT threads into fabric is the periodic patterning of CNT threads into p–n regions. For periodic n-doping of the p-type CNT thread, part of the CNT thread was dipped into 1-butyl-3-methylimidazolium hexafluorophosphate ($[BMIM]PF_6$) used as an n-dopant to create a p–n striped pattern. The stripe-patterned thread was sewn into a felt fabric (ca. 3 mm in thickness) with a sewing needle such that the n-type and p-type sections traversed the fabric in the downward and upward directions, respectively. The TE fabric thus obtained exhibited ca. 8 nW at a $\Delta T = 25$ K. In this 3D π-shaped structure, thick felt functioned as an efficient thermal insulator, allowing effective utilization of the temperature gradient.

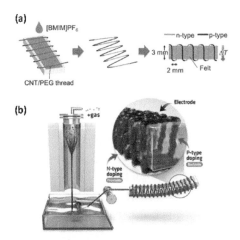

Figure 6.17. *(a) Schematic illustration of the fabrication process of a TE fabric. Reproduced with permission (Ito et al. 2017). Copyright 2017, The Royal Society of Chemistry. (b) Schematic illustration of the flexible TEG based on CNT yarn. Reproduced with permission (Choi et al. 2017). Copyright 2017, American Chemical Society*

Choi et al. reported a novel patterning method using a CNT yarn, and the patterned yarn was used as a 3D π-shaped TEG (see Figure 6.17(b)) (Choi et al. 2017). The CNT yarns were synthesized by the floating catalyst method and directly spun around a PDMS rectangular support. The CNT yarns on one side of the support were doped with $FeCl_3$ for p-doping, and that on the front side were doped with PEI solutions for n-doping. The CNT yarns on the other two sides were left undoped as electrodes. The obtained 3D π-shaped TEG showed an exceptional power density of 697 µW·g^{-1} at a ΔT of 40 K with 60 couples of p- and n-type CNT yarns, primarily because of the extremely high power factor of the p-type (2,387 µW m^{-1} K^{-2}) and n-type (2,456 µW m^{-1} K^{-2}) regions. This research strongly suggests an advantage of the CNT yarn directly spun from the synthesis furnace. Additionally, the patterning method is also useful for CNT-based TEG fabrics because of their strong scalability.

Recently, several attempts have been made to fabricate large-area CNT-based TE fabrics. Zheng et al. developed a simple process to prepare p–n patterns for CNT yarns (Zheng et al. 2020). In this process, the CNT yarn is wrapped onto a thin polyethylene terephthalate (PET) plate and rolled into a cylindrical structure (see Figure 6.18). Subsequently, the CNT yarn on the PET roll is soaked in a PEDOT:PSS solution with a controlled dipping time, and the roll is flipped over and immersed into a PEI/ethanol solution with a certain doping time. After reinforcing with the PET fiber, the patterned CNT yarn is sewn into the warp-knitted spacer fabric using a needle. The optimally designed TEG with good wearability and stability shows a high output power density of 5.15 µW cm^{-2} and an extremely high specific power of 244.6 µW g^{-1} at ΔT of 47.5 K. In addition to their high power factor of the CNT/PEDOT:PSS composite yarn (512.8 µW mK^{-2}) as well as the PEI/CNT yarn (667.8 µW mK^{-2}), sophisticated design to integrate CNT yarns into TE fabric offers high power density.

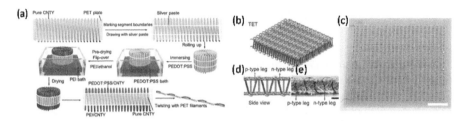

Figure 6.18. *(a) Schematic illustration of the patterning process of CNT yarns. (b) Diagrams and (c) optical images of the as-prepared warp-knitted TEG. (d) Schematic diagram and (e) image at the bottom of the cross-section of the as-prepared TEB, demonstrating the alternatively vertical arrangement of p-type and n-type legs (scale bar: 2 mm). Reproduced with permission (Zheng et al. 2020). Copyright 2020, The Royal Society of Chemistry*

In 2020, Sun et al. reported a π-shaped CNT-based TE fabric that is directly woven into textiles, instead of embedding it into clothes using CNT fibers prepared by twisting CNT films (Sun et al. 2020). To create p–n patterns, p-doping was carried out by dipping into PEDOT:PSS solution and n-doping was performed with an electrospray of n-dopant solution using polymer as a mask. It is important to note that the electrospray doping method generates monodispersed microdroplets containing a doping solution with repulsive electric charges. They proved that this method is superior to the conventional dipping method, which has a low patterning resolution due to diffusion. TE fibers with alternatively doped n- and p-segments were wrapped with acrylic fibers to avoid a short circuit (see Figure 6.19). Patterned CNT fibers were used to fabricate alternately interlocked TE loops that offer thermal resistance matching as well as excellent stretchability. The

TEG textile composed of interlocked TE loops shows a peak power density of 7.0 µWcm^{-2} for a ΔT of 44 K.

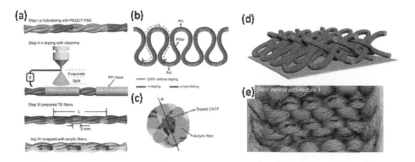

Figure 6.19. *(a) Patterning of CNT fiber using an electrospray technique, followed by wrapping with acrylic fibers. (b) Loop structure of TE leg. (c) Cross-sectional view of the structure of the CNT fiber wrapped by acrylic fibers. (d) 3D interlocked TEGs without substrate. Reproduced with permission (Sun et al. 2020). Copyright 2020, Nature Publishing Group*

Importantly, they compared their performance in terms of specific power density, in which the power density was normalized to the temperature difference squared ΔT^2 because the power is proportional to the squared temperature difference. Such a comparison is reasonable when considering the diversity of the given temperature differences studied in different reports. As a result, the output performance of the TEG was 35 µWm^{-2}K^{-2} with a 16 mm repeat length, and this value was superior to that of other flexible CNT TEGs (Mai et al. 2015; Kim et al. 2014; Hewitt et al. 2012; Toshima et al. 2015; Choi et al. 2016; Ito et al. 2017; Wu et al. 2017; Zhou et al. 2017).

6.9. Conclusion

In this chapter, current understandings of TE properties of CNTs and their control methods mainly based on the chemical doping method are summarized. Especially, large progress in the TE studies for semiconducting SWCNT is highlighted. Also, the progress of n-doping technology in these 10 years is discussed. Developments of the fabrication and purification technology of CNT will offer further progress of the TE properties. In addition, the history of the developments of TEGs based on CNTs is summarized in terms of the TEG structure. Starting from the 2D π-shape configuration, recent developments have focused on the 3D π-shaped structure to use the temperature gradient perpendicular to the surface. In particular, the embedding of the π-shaped structure into a fabric using CNT fiber has attracted significant attention owing to its low density and compatibility with wearable applications.

6.10. References

An, C.J., Kang, Y.H., Song, H., Jeong, Y., Cho, S.Y. (2017). High-performance flexible thermoelectric generator by control of electronic structure of directly spun carbon nanotube webs with various molecular dopants. *J. Mater. Chem. A*, 5, 15631–15639.

Avery, A.D., Zhou, B.H., Lee, J., Lee, E.-S., Miller, E.M., Ihly, R., Wesenberg, D., Mistry, K.S., Guillot, S.L., Zink, B.L. et al. (2016). Tailored semiconducting carbon nanotube networks with enhanced thermoelectric properties. *Nature Energy*, 1, 16033.

Bahk, J.-H., Fang, H., Yazawa, K., Shakouri, A. (2015). Flexible thermoelectric materials and device optimization for wearable energy harvesting. *J. Mater. Chem. C*, 3, 10362–10374.

Blackburn, J.L., Ferguson, A.J., Cho, C., Grunlan, J.C. (2018). Carbon-nanotube-based thermoelectric materials and devices. *Adv. Mater.*, 30, 1704386.

Bradley, K., Jhi, S.-H., Collins, P.G., Hone, J., Cohen, M.L., Louie, S.G., Zettl, A. (2000). Is the intrinsic thermoelectric power of carbon nanotubes positive? *Phys. Rev. Lett.*, 85, 4361–4364.

Cho, C., Stevens, B., Hsu, J.-H., Bureau, R., Hagen, D.A., Regev, O., Yu, C., Grunlan, J.C. (2015). Completely organic multilayer thin film with thermoelectric power factor rivaling inorganic tellurides. *Adv. Mater.*, 27, 2996–3001.

Cho, C., Wallace, K.L., Tzeng, P., Hsu, J.H., Yu, C., Grunlan, J.C. (2016). Outstanding low temperature thermoelectric power factor from completely organic thin films enabled by multidimensional conjugated nanomaterials. *Adv. Energy Mater.*, 6.

Choi, J., Lee, J.Y., Lee, S.S., Park, C.R., Kim, H. (2016). High-performance thermoelectric paper based on double carrier-filtering processes at nanowire heterojunctions. *Adv. Energy Mater.*, 6.

Choi, J., Jung, Y., Yang, S.J., Oh, J.Y., Oh, J., Jo, K., Son, J.G., Moon, S.E., Park, C.R., Kim, H. (2017). Flexible and robust thermoelectric generators based on all-carbon nanotube yarn without metal electrodes. *ACS Nano*, 11, 7608–7614.

Dörling, B., Ryan, J.D., Craddock, J.D., Sorrentino, A., Basaty, A.E., Gomez, A., Garriga, M., Pereiro, E., Anthony, J.E., Weisenberger, M.C. et al. (2016). Photoinduced p- to n-type switching in thermoelectric polymer-carbon nanotube composites. *Adv. Mater.*, 28, 2782–2789.

Hewitt, C.A., Kaiser, A.B., Roth, S., Craps, M., Czerw, R., Carroll, D.L. (2012). Multilayered carbon nanotube/polymer composite based thermoelectric fabrics. *Nano Lett.*, 12, 1307–1310.

Hicks, L.D. and Dresselhaus, M.S. (1993). Thermoelectric figure of merit of a one-dimensional conductor. *Phys. Rev. B*, 47, 16631–16634.

Hone, J., Ellwood, I., Muno, M., Mizel, A., Cohen, M.L., Zettl, A., Rinzler, A.G., Smalley, R.E. (1998). Thermoelectric power of single-walled carbon nanotubes. *Phys. Rev. Lett.*, 80, 1042–1045.

Hong, C.T., Kang, Y.H., Ryu, J., Cho, S.Y., Jang, K.-S. (2015). Spray-printed CNT/P3HT organic thermoelectric films and power generators. *J. Mater. Chem. A*, 3, 21428–21433.

Ito, M., Koizumi, T., Kojima, H., Saito, T., Nakamura, M. (2017). From materials to device design of a thermoelectric fabric for wearable energy harvesters. *J. Mater. Chem. A*, 5, 12068–12072.

Kim, D., Kim, Y., Choi, K., Grunlan, J.C., Yu, C.H. (2010). Improved thermoelectric behavior of nanotube-filled polymer composites with poly(3,4-ethylenedioxythiophene) poly(styrenesulfonate). *Acs Nano*, 4, 513–523.

Kim, S.L., Choi, K., Tazebay, A., Yu, C. (2014). Flexible power fabrics made of carbon nanotubes for harvesting thermoelectricity. *ACS Nano*, 8, 2377–2386.

Kim, J.-Y., Lee, W., Kang, Y.H., Cho, S.Y., Jang, K.-S. (2018). Wet-spinning and post-treatment of CNT/PEDOT:PSS composites for use in organic fiber-based thermoelectric generators. *Carbon*, 133, 293–299.

Kim, S., Mo, J.-H., Jang, K.-S. (2019). Solution-processed carbon nanotube buckypapers for foldable thermoelectric generators. *ACS Appl. Mater. Interfaces*, 11, 35675–35682.

Lee, M.-H., Kang, Y.H., Kim, J., Lee, Y.K., Cho, S.Y. (2019). Freely shapable and 3D porous carbon nanotube foam using rapid solvent evaporation method for flexible thermoelectric power generators. *Adv. Energy Mater.*, 9, 1900914.

Macleod, B.A., Stanton, N.J., Gould, I.E., Wesenberg, D., Ihly, R., Owczarczyk, Z.R., Hurst, K.E., Fewox, C.S., Folmar, C.N., Hughes, K.H. et al. (2017). Large n-and p-type thermoelectric power factors from doped semiconducting single-walled carbon nanotube thin films. *Energy Environ. Sci.*, 10, 2168–2179.

Mai, C.-K., Russ, B., Fronk, S.L., Hu, N., Chan-Park, M.B., Urban, J.J., Segalman, R.A., Chabinyc, M.L., Bazan, G.C. (2015). Varying the ionic functionalities of conjugated polyelectrolytes leads to both p- and n-type carbon nanotube composites for flexible thermoelectrics. *Energy Environ. Sci*, 8, 2341–2346.

Mytafides, C.K., Tzounis, L., Karalis, G., Formanek, P., Paipetis, A.S. (2021). High-power all-carbon fully printed and wearable SWCNT-based organic thermoelectric generator. *ACS Appl. Mater. Interfaces*, 13, 11151–11165.

Nakai, Y., Honda, K., Yanagi, K., Kataura, K., Kato, T., Yamamoto, T., Maniwa, Y. (2014). Giant seebeck coefficient in semiconducting single-wall carbon nanotube film. *Appl. Phys. Express*, 7, 025103.

Nakano, M., Nakashima, T., Kawai, T., Nonoguchi, Y. (2017). Synergistic impacts of electrolyte adsorption on the thermoelectric properties of single-walled carbon nanotubes. *Small*, 13.

Nakashima, Y., Nakashima, N., Fujigaya, T. (2017). Development of air-stable n-type single-walled carbon nanotubes by doping with 2-(2-methoxyphenyl)-1, 3-dimethyl-2,3-dihydro-1H-benzo[d]imidazole and their thermoelectric properties. *Synth. Met.*, 225, 76–80.

Nakashima, Y., Yamaguchi, R., Toshimitsu, F., Matsumoto, M., Borah, A., Staykov, A., Islam, M.S., Hayami, S., Fujigaya, T. (2019). Air-stable n-type single-walled carbon nanotubes doped with benzimidazole derivatives for thermoelectric conversion and their air-stable mechanism. *ACS Appl. Nano Mater.*, 2, 4703–4710.

Nan, K., Kang, S.D., Li, K., Yu, K.J., Zhu, F., Wang, J., Dunn, A.C., Zhou, C., Xie, Z., Agne, M.T. et al. (2018). Compliant and stretchable thermoelectric coils for energy harvesting in miniature flexible devices. *Sci. Adv.*, 4, eaau5849.

Nandihalli, N., Liu, C.-J., Mori, T. (2020). Polymer based thermoelectric nanocomposite materials and devices: Fabrication and characteristics. *Nano Energy*, 78, 105186.

Nonoguchi, Y., Ohashi, K., Kanazawa, R., Ashiba, K., Hata, K., Nakagawa, T., Adachi, C., Tanase, T., Kawai, T. (2013). Systematic conversion of single walled carbon nanotubes into n-type thermoelectric materials by molecular dopants. *Sci. Rep.*, 3.

Nonoguchi, Y., Sudo, S., Tani, A., Murayama, T., Nishiyama, Y., Uda, R.M., Kawai, T. (2017a). Solvent basicity promotes the hydride-mediated electron transfer doping of carbon nanotubes. *Chem. Commun.*, 53, 10259–10262.

Nonoguchi, Y., Tani, A., Ikeda, T., Goto, C., Tanifuji, N., Uda, R.M., Kawai, T. (2017b). Water-processable, air-stable organic nanoparticle–carbon nanotube nanocomposites exhibiting n-type thermoelectric properties. *Small*, 13, 1603420.

Nonoguchi, Y., Kojiyama, K., Kawai, T. (2018). Electrochemical n-type doping of carbon nanotube films by using supramolecular electrolytes. *J. Mater. Chem. A*, 6, 21896–21900.

Park, K.T., Choi, J., Lee, B., Ko, Y., Jo, K., Lee, Y.M., Lim, J.A., Park, C.R., Kim, H. (2018). High-performance thermoelectric bracelet based on carbon nanotube ink printed directly onto a flexible cable. *J. Mater. Chem. A*, 6, 19727–19734.

Shim, M., Javey, A., Shi Kam, N.W., Dai, H. (2001). Polymer functionalization for air-stable n-type carbon nanotube field-effect transistors. *J. Am. Chem. Soc.*, 123, 11512–11513.

Small, J.P., Perez, K.M., Kim, P. (2003). Modulation of thermoelectric power of individual carbon nanotubes. *Phys. Rev. Lett.*, 91, 256801.

Suemori, K., Hoshino, S., Kamata, T. (2013). Flexible and lightweight thermoelectric generators composed of carbon nanotube–polystyrene composites printed on film substrate. *Appl. Phys. Lett.*, 103, 153902.

Sun, T., Zhou, B., Zheng, Q., Wang, L., Jiang, W., Snyder, G.J. (2020). Stretchable fabric generates electric power from woven thermoelectric fibers. *Nat. Commun.*, 11, 1–10.

Tanaka, N., Hamasuna, A., Uchida, T., Yamaguchi, R., Ishii, T., Staylkov, A., Fujigaya, T. (2021). Electron doping of single-walled carbon nanotubes using pyridine-boryl radicals. *Chem. Commun.*, 57, 6019–6022.

Toshima, N., Oshima, K., Anno, H., Nishinaka, T., Ichikawa, S., Iwata, A., Shiraishi, Y. (2015). Novel hybrid organic thermoelectric materials: Three-component hybrid films consisting of a nanoparticle polymer complex, carbon nanotubes, and vinyl polymer. *Adv. Mater.*, 27, 2246–2251.

Wei, Q., Mukaida, M., Kirihara, K., Ishida, T. (2014). Experimental studies on the anisotropic thermoelectric properties of conducting polymer films. *ACS Macro Letters*, 3, 948–952.

Wu, G., Zhang, Z.-G., Li, Y., Gao, C., Wang, X., Chen, G. (2017). Exploring high-performance n-type thermoelectric composites using amino-substituted rylene dimides and carbon nanotubes. *ACS Nano*, 11, 5746–5752.

Yamaguchi, R., Ishii, T., Matsumoto, M., Borah, A., Tanaka, N., Oda, K., Tomita, M., Watanabe, T., Fujigaya, T. (2021). Thermal deposition method for p–n patterning of carbon nanotube sheets for planar-type thermoelectric generator. *J. Mater. Chem. A*, 9, 12188–12195.

Yanagi, K., Kanda, S., Oshima, Y., Kitamura, Y., Kawai, H., Yamamoto, T., Takenobu, T., Nakai, Y., Maniwa, Y. (2014). Tuning of the thermoelectric properties of one-dimensional material networks by electric double layer techniques using ionic liquids. *Nano Lett.*, 14, 6437–6442.

Yao, Q., Chen, L., Zhang, W., Liufu, S., Chen, X. (2010). Enhanced thermoelectric performance of single-walled carbon nanotubes/polyaniline hybrid nanocomposites. *ACS Nano*, 4, 2445–2451.

Yoshiyuki, N., Motohiro, N., Tomoko, M., Harutoshi, H., Shota, H., Koji, M., Ryosuke, M., Masakazu, N., Tsuyoshi, K. (2016). Simple salt-coordinated n-type nanocarbon materials stable in air. *Adv. Funct. Mater.*, 26, 3021–3028.

Yu, C., Kim, Y.S., Kim, D., Grunlan, J.C. (2008). Thermoelectric behavior of segregated-network polymer nanocomposites. *Nano Lett.*, 8, 4428–4432.

Yu, C., Murali, A., Choi, K., Ryu, Y. (2012). Air-stable fabric thermoelectric modules made of N- and P-type carbon nanotubes. *Energy Environ. Sci.*, 5, 9481–9486.

Zheng, Y., Zhang, Q., Jin, W., Jing, Y., Chen, X., Han, X., Bao, Q., Liu, Y., Wang, X., Wang, S. et al. (2020). Carbon nanotube yarn based thermoelectric textiles for harvesting thermal energy and powering electronics. *J. Mater. Chem. A*, 8, 2984–2994.

Zhou, W., Fan, Q., Zhang, Q., Cai, L., Li, K., Gu, X., Yang, F., Zhang, N., Wang, Y., Liu, H., Zhou, W. et al. (2017). High-performance and compact-designed flexible thermoelectric modules enabled by a reticulate carbon nanotube architecture. *Nat. Commun.*, 8, 14886.

PART 3

Metrology of Thermal Properties

7

Precise Measurement of the Absolute Seebeck Coefficient from the Thomson Effect

Yasutaka AMAGAI
National Metrology Institute of Japan, National Institute of Advanced Industrial Science and Technology (AIST), Japan

7.1. Introduction

In the middle of the 19th century, William Thomson, known as Lord Kelvin, discovered that heat is released or absorbed in proportion to an amplitude of electrical current and temperature difference in a conductor (Thomson 1851). In the original experimental setup shown in Figure 7.1, an electric current passed through a U-shaped Fe rod. Two electrical resistance coils, R_1 and R_2, were wound to form a detector and were connected to the Wheatstone bridge. The bottom of the U-shaped Fe rod was then heated, which yielded two temperature gradients, a positive one extending from A to C, and a negative one extending from C to B (Duckworth 1960). This thermoelectric effect is referred to as the Thomson effect. When a charge current density j and a temperature gradient dT/dx are applied to a conductor, the heat production rate per unit volume may be described as

$$q = \frac{j^2}{\sigma} - \mu j \frac{dT}{dx}, \qquad [7.1]$$

where μ is defined as the Thomson coefficient, and σ is the electrical conductivity. The first term is the thermodynamically irreversible Joule heat, and the second term is the Thomson heat. The Thomson effect is the only thermoelectric effect that can occur in a single and homogeneous conductor unlike the Seebeck effect and Peltier effect (Seebeck 1825; Peltier 1834). Therefore, combining the thermoelectric Kelvin relation (Thomson 1854), the measurement of the Thomson coefficient μ provides the means of experimentally determining the absolute Seebeck coefficient known as an absolute scale of thermoelectricity.

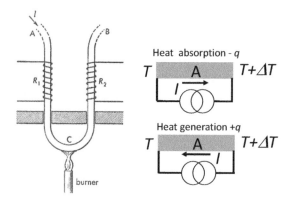

Figure 7.1. *Drawing of the original experimental setup for the measurement of the Thomson effect. Reprinted from Duckworth (1960, Figure 7.4, p. 183). The schematic illustration of the Thomson effect that occurred in a single and homogeneous conductor is also shown on the right*

With recent advancements in thermophysical property metrology, such as the development of new thermoelectric standard reference materials (SRMs) (Lowhorn et al. 2009; Lenz et al. 2013; Martin et al. 2021), and international comparisons (Wang et al. 2013; Alleno et al. 2015) – it may be helpful to revisit the fundamentals of the measurement methods pertaining to the Thomson effect. We note that there are excellent general reviews (Martin et al. 2010; Borup et al. 2015; Akinaga et al. 2020) and books (Macdonald 1962; Blatt and Schroeder 1978; Rowe 1995; Nolas et al. 2001; Tritt 2004) on the fundamentals of thermoelectricity and measurements. However, there are few reports specifically addressing Thomson heat measurements, and fewer regarding the metrological application to the absolute scale of thermoelectricity.

The purpose of this chapter is to introduce a fundamental concept of Thomson heat measurements and the application to the absolute scale of thermoelectricity. We begin with a brief review of the measurement principle of the absolute Seebeck coefficient. General concepts such as the Kelvin relation in the thermoelectric effect are introduced at a conceptual level. The content is then organized in the measurement methods of Thomson heat measurements. A description of the conventional method and the recently developed AC–DC method, along with a 2ω technique, and a lock-in thermal imaging technique, are provided with an outlook and summary.

7.2. Absolute scale of thermoelectricity

The Seebeck coefficient S, also known as thermoelectric power or thermopower, is defined as the ratio of the induced thermoelectric voltage ΔV to the applied temperature difference ΔT in a conductor:

$$S \equiv \lim_{\Delta T \to 0} \frac{\Delta V}{\Delta T}. \qquad [7.2]$$

The Seebeck coefficient appears to be conceptually simple; however, it is actually cumbersome to determine the absolute value accurately. As shown in Figure 7.2, a thermoelectric voltage V_{AB} experimentally observed in the open-circuit condition can be expressed as (Ziman 1972)

$$V_{AB} = \int_T^{T+\Delta T} (S_B - S_A) dT = (S_B - S_A) \Delta T. \qquad [7.3]$$

Rearranging equation [7.3], we obtain

$$S_A = S_B + \frac{V_{AB}}{\Delta T}, \qquad [7.4]$$

where T and $T + \Delta T$ are the temperatures at the edges of the sample. S_A and S_B are the Seebeck coefficients of the conductor A (sample) and the conductor B (two voltage leads), respectively. This circuit analysis indicates that while the Seebeck coefficient can be clearly defined for a single and homogeneous conductor, the measured thermoelectric voltage in popular thermocouple experiments is always proportional to the difference between the Seebeck coefficient of the sample of interest (A) and that of the reference voltage leads (B). Therefore, the Seebeck coefficient of the reference voltage leads must be determined with a separate experiment.

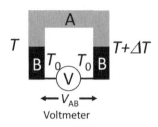

Figure 7.2. *The schematic illustration of the Seebeck effect. The sample of interest (A) is connected with the voltmeter through the voltage leads (B)*

Superconductors can be used as Seebeck reference materials in low-temperature regions, as they essentially have a zero Seebeck coefficient well below their transition temperature T_c. This intrinsic superconductor property occurs because electric currents, induced by the temperature gradient in the superconductor, are canceled by supercurrent counterflows in accordance with the Meissner effect (Ginzburg 1990). Thus, by making one of the wires (see conductor B in Figure 7.2) in a thermocouple from a superconducting material, it is possible to directly measure the absolute Seebeck coefficient. In particular, high-T_c superconductors, such as $YBa_2Cu_3O_{7-x}$ (Uher 1987) and $Bi_2Sr_2Ca_2Cu_3O_{8+\delta}$ (Amagai et al. 2020), may be useful because they could offer a wide operating temperature range above liquid nitrogen temperature.

Similarly, the Thomson coefficient measurement is indispensable as a means of experimentally determining the absolute Seebeck coefficient of a reference material at high temperatures (above approximately 100 K), where superconducting reference materials are no longer used. Originally proposed by William Thomson, known as Lord Kelvin, the above three thermoelectric effects are combined with Kelvin's relation (Thomson 1854). Kelvin's relation is a specific case of Onsager's reciprocal relations (Onserger 1931a, 1931b; Callen 1948). As such, this relation involves microscopic reversibility, a fundamental principle in physics. The Thomson effect links the Peltier coefficient Π and the Seebeck coefficient S at any temperature as follows:

$$\frac{d\Pi}{dT} = S + \mu,$$

$$\Pi = ST.$$
[7.5]

The Kelvin relation was verified for several thermocouple pairs of metals measured on the same sample. A large amount of data was compiled, and the relation between S and π/T appearing in the second relation in equation [7.5] was reasonably close to the theoretically predicted value obtained from Onsager's reciprocal relations within the measurement uncertainty. Thus, the validity of the Kelvin relation, a special case of Onserger's reciprocal ratio, can be accepted with considerable confidence (Miller 1960).

Combining Kelvin's relations, the Thomson coefficient μ can be related to the Seebeck coefficient through temperature variation

$$\mu = T \frac{dS}{dT}$$
[7.6]

Integrating the above relation with respect to the temperature T, we can determine the absolute Seebeck coefficient from calorimetric measurements of the Thomson heat

$$S(T_1) - S(0) = \int_0^{T_1} \frac{\mu}{T} dT,$$
[7.7]

where T_1 is the temperature of interest. According to the third law of thermodynamics, the Seebeck coefficient becomes zero as the absolute temperature approaches zero. Thus, we obtain the following equation:

$$S(T_1) = \int_0^{T_1} \frac{\mu}{T} dT$$
[7.8]

In this manner, an absolute scale of thermoelectricity can be established at high temperatures. Historically, Pb has long been used as a Seebeck reference material because of its small Seebeck coefficient (less than 1 µV/V) and relatively high superconducting transition temperature of approximately 7 K. In the early 20th century, the first absolute scale of thermoelectricity for Pb was established by superconducting thermocouple experiments, in which one couple consists of superconducting material, and from the indirect measurement of the Thomson heat of Pb (Borelius et al. 1932). To reevaluate the scale, Nb_3Sn, which has a higher T_C of 23 K, was employed in a thermocouple experiment as a reference from 18 K to 20 K, and the scale was calculated by the superconducting experiments using Nb_3Sn as a reference, and Borelius' values obtained by the indirect measurement of the Thomson coefficient (Christian et al. 1958). Although Pb seems an appropriate choice in this context, its practical use as a reference material is limited because of the difficulty in fabricating sufficiently thin wires, in addition to the toxic properties of this element. Instead, Pt and Cu wires are routinely employed. Subsequently, the scale was extended up to 1,300 K for Cu and 2,000 K for Pt by compiling the available data (Cusack and Kendall 1958). Direct thermocouple experiments for Pt using a Pb wire as a reference from 80 K to 400 K were also performed to recalculate Cusack's high-temperature scale in the 1970s (Moore and Graves 1973). However, a direct measurement of the Thomson coefficient of Pb has not been made since the establishment of the absolute scale of thermoelectricity by Borelius. We note that Borelius's values for the Thomson coefficient were obtained indirectly from measurements of the Thomson heat of an alloy and subsequent thermocouple experiments relative to the same alloy. For this reason, the direct measurement of the Thomson coefficient of high-purity Pb was performed over a complete temperature range of 4–300 K (Roberts 1977a). The values for μ agree with those from the superconducting thermocouple experiment by Christian et al. from 10 K to 17 K, and with those from the indirect measurements of Borelius et al. from 80 K to 300 K. However, Roberts found that from 17 K to 80 K, disagreement with the values used by Christian et al. to construct their absolute scale of thermoelectricity leads to a significant change in the scale of approximately 0.3 μV/K from 30 K to 300 K, corresponding to a relative error of 30%.

Today, the modern standard was established by Roberts' results for Pb from 10 K to 300 K (Roberts 1977b) for Cu between 273 K and 900 K, for Pt between 273 K and 1,600 K and for W between 273 K and 1,800 K, obtained from direct measurements of the Thomson coefficient (Roberts 1981; Roberts et al. 1985). The measurement uncertainty at the highest temperatures for Pt was estimated to be 0.2 μV/K. For Pb in the range from 10 K to 300 K, the measurement uncertainty was estimated to be 0.01 μV/K. Recently, Burkov provided an empirical interpolation function for the Seebeck coefficient of Pt between 70 K and 1,500 K according to the new absolute scale (Burkov et al. 2001). The present absolute scales for Pb, Pt, Cu and W are shown in Figure 7.3(a). The enlarged view of the Seebeck coefficient in Pb in Figure 7.3(b) indicates the discrepancy between the previous scale by Christian and the present scale by Roberts. The unexpected change near 20 K causes a great difference in the shape of the calculated Seebeck coefficients of Pb between the two scales, as Seebeck coefficients are as small as 1 µV/K. These results have served as the ultimate foundation for nearly all subsequent thermocouple experiments performed to determine the Seebeck coefficient, as well as continuing theoretical research over the half century.

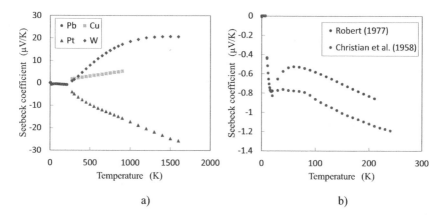

Figure 7.3. *Present absolute scale of thermoelectricity determined from the direct measurement of the Thomson effect in Pb. (a) The absolute scales for Pb, Pt, Cu and W and (b) enlarged view of the absolute Seebeck coefficient of Pb obtained from the direct Thomson heat measurement by Roberts (1977b) and indirect measurement by Christian et al. (1958). For a color version of this figure, see www.iste.co.uk/akinaga/thermoelectric1.zip*

7.3. Measurement methods of the Thomson effect

Accurate measurements of the Thomson coefficient are typically expensive and time intensive, in part because of the difficulty in isolating a single sample and measuring the low Thomson heat. Early accurate measurements of the Thomson effect were conducted for several pure metals by measuring the temperature profile of the conductors instead of measuring the heat flow caused by the Thomson effect (Nettleton 1912, 1921; Lander 1948). The Thomson effect at low temperatures has received considerable attention from theoretical perspectives, such as in considering the Seebeck effect in alkaline and transition metals (Pearson and Templeton 1958; Macdonald 1962; Maxwell et al. 1967). In the establishment of the absolute scale, a differential thermocouple technique was employed to measure the temperature difference between two parallel insulated wires subject to the same temperature gradient, but with DC current flowing through them in opposite directions (Roberts 1977b). However, until now, there has been little conceptual development of new methods since early developments by Nettleton (1912). Recently, new methods that employ AC and DC current signals have been developed: an AC–DC method (Amagai et al. 2015, 2019a, 2020), a 2ω technique (Dunn et al. 2019) and a thermoelectric imaging technique (Uchida et al. 2020). At sufficiently high frequencies, the Thomson effect can be canceled out, which allows the small Thomson heat to be extracted from the large Joule heat. Most importantly, the use of both AC and DC currents significantly simplifies the expression of the Thomson coefficient and compensates for the heat loss from the sample to its surroundings. Meanwhile, the 2ω technique is a phase-sensitive method that uses the response at the second harmonic of the excitation frequency of the electrical resistance oscillation in the sample, which is directly proportional to the Thomson coefficient. The phase-sensitive method has a distinctive advantage over a DC detection technique, because a small AC voltage can be extracted from high background noise. The thermoelectric imaging technique uses a non-contact lock-in thermography technique to detect the small Thomson heat. This method is proven to be a powerful tool for evaluating the magnetic dependence of the Thomson effect and may be applicable for the determination of the Thomson coefficient.

7.3.1. Conventional method

If a DC current I passes through the sample along the temperature gradient, the measured temperature change of the sample is attributed to two main factors. The first is Joule heating, which is proportional to the square of the current amplitude. The second is the Thomson effect, which is proportional to the current amplitude. As Joule heating occurs, the Thomson effect is changed from heating to cooling by reversing the current through the sample. Thus, the contribution of the Joule effect can be eliminated by averaging both temperatures. Then, applying some algebra directly produces the Thomson coefficient (Nettleton 1912):

$$\mu = \frac{8K\Delta T_{\text{DC}}}{\Delta T I}. \qquad [7.9]$$

Here, K is the thermal conductance, defined as $K \equiv \kappa a/l$, where a and l are the cross-sectional area and length of the sample, respectively, and ΔT represents the temperature difference across the sample. Moreover, ΔT_{DC} caused by the Thomson effect is calculated from $\Delta T_{DC} \equiv (T_{+DC} - T_{-DC})/2$. In addition, T_{+DC} and T_{-DC} correspond to the temperature at the middle of the sample when positive and negative DC currents are applied, respectively. Experimentally, the Thomson coefficient, given by equation [7.9], was measured using the configuration depicted in Figure 7.4 (Nettleton 1921). The sample was mechanically clamped to a wound heat reservoir to yield a temperature gradient along the sample. The temperature was applied to the cylindrical Cu heat sink. The surrounding temperature of the sample was maintained at the same temperature as that of the cold side of the reservoir. A detecting coil made from insulated Cu was attached to the middle of the sample. This innovative technique allows accurate determination of the Thomson coefficient and has been employed in many experiments; however, we note that equation [7.9] was derived in an ideal isothermal condition, where the heat loss from the sample is disregarded. Furthermore, prior information regarding the thermal conductivity value of the sample and the sample's dimensions is required, which would make the measurements time intensive.

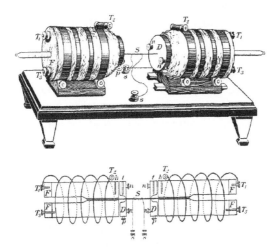

Figure 7.4. *Original experimental setup for the measurement of the Thomson effect according to Nettleton's formula [7.9]. Reproduced from Nettleton (1921), with permission. All rights reserved*

7.3.2. *New measurement methods: AC–DC method*

The full amount of heat created by the Thomson effect or Joule effect is used to heat or cool the sample, while disregarding the heat loss from the sample. Accordingly, the temperature profile of the sample is $T-T_0$, where T_0 is the environmental temperature, normalized by the temperature difference of the sample ΔT, defined as $T_H - T_C$, and will resemble the black solid curves in Figure 7.5(a) in the absence of radiative heat loss. Here, T_h and T_c are hot and cold sides of the temperature of the sample, respectively. Black solid curves were obtained for $\mu I/K = 0, \pm 3$. The temperature increases at the middle of the sample, corresponding to $x/l = 0.5$, induced by applying a positive and negative DC current and is measured, and equation [7.10] produces an accurate Thomson coefficient. In practice, however, the sample will exchange heat with the environment as radiative heat. Correspondingly, the temperature profiles will resemble the lower dotted black curves in Figure 7.5(a). Consequently, equation [7.9] will erroneously yield a Thomson coefficient that is too low, even if the other parameters such as the thermal conductivity or sample geometry are measured accurately. In the measurement configuration shown in Figure 7.5(b), simply adding an AC current source to a conventional configuration solves this problem completely by accounting for the temperature increase ΔT_{AC} caused by an AC current that is equivalent to the DC current I: AC–DC method. The underlying concept of the AC–DC method is that most of the radiative heat loss occurring during the measurement of ΔT_{AC} and ΔT_{DC} can be compensated for by calculating the ratio of the two signals. If we now consider the radiative coupling to the environment by simulating the temperature profile based on the heat transfer equation, the Thomson coefficient in equation [7.9] can be expanded to the leading order using χ:

$$\mu = \frac{8K\Delta T_{DC}}{\Delta T I}\left[1 + \chi + \frac{5}{6}\chi^2 + O(\chi^3) \right] \qquad [7.10]$$

where χ is defined as $\chi \equiv (p\gamma l)/8K$. Here, γ is defined as $\gamma = 4\sigma\varepsilon T_0^3$, where σ is the Stefan–Boltzmann constant, ε is the emissivity and T_0 is the surrounding temperature. In addition, p in the fourth term denotes the sample perimeter. The expanded Thomson coefficient in equation [7.10] includes the thermal conductivity, sample geometry and correction term related to the sample heat loss due to radiative heat transfer. In contrast, upon rearranging equation [7.10], substituting the temperature change ΔT_{AC} caused by the Joule effect at the middle of the sample when the AC current equivalent to the DC current is applied, and setting the electrical resistance of the sample as $R = 2\rho l/a$, the Thomson coefficient can be modified to

$$\mu = \frac{IR}{\Delta T}\frac{\Delta T_{DC}}{\Delta T_{AC}}\left[1+\frac{1}{6}\chi+\frac{7}{45}\chi^2+O(\chi^3)\right]$$ [7.11]

$$\approx \frac{IR}{\Delta T}\frac{\Delta T_{DC}}{\Delta T_{AC}}.$$ [7.12]

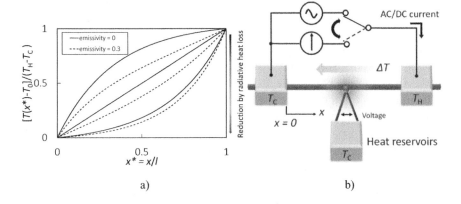

Figure 7.5. (a) Influence of radiative heat loss on the temperature profiles along the sample caused by the Thomson effect for emissivity $\varepsilon = 0$ and $\varepsilon = 0.3$, where the thermal conductivity is 71.6 W/mK and the diameter is 500 μm. Here, l = 138 mm, and the resistivity $\rho = 1.06 \times 10^{-8}$ Ωm. To show how Thomson heating or cooling affects the temperature distribution, three parameters, i.e. $\mu I/K$ = 0, ± 3 are chosen for each emissivity, where K is the heat conductance of the sample, defined as $K \equiv \kappa a/l$. (b) Schematic of the modified configuration for the Thomson coefficient measurement. The temperatures at $x = 0$ and $x = l$ are fixed at T_C and T_H, respectively, and the thermocouple is placed in the middle of the sample. Reproduced from Amagai et al. (2019b), with the permission of AIP Publishing. For a color version of this figure, see www.iste.co.uk/akinaga/thermoelectric1.zip

We note that the first correction term appearing in equation [7.10] is reduced by a factor of 6 upon calculating the ratio of the temperature changes ΔT_{DC} and ΔT_{AC} in equation [7.11]. Therefore, compensating for radiative heat loss requires only two additional measurements: Joule heat measurement and electrical resistance measurement. These measurements require much less effort than other approaches, such as the use of multiple buffers or accurate temperature control of thermal shields. Most importantly, the proposed formula in equation [7.12] does not require prior knowledge of the thermal conductivity and emissivity of the sample, unlike the conventional formula [7.9].

The uncompensated Thomson coefficient values of Pt, obtained directly from equation [7.9], together with the compensated values using equation [7.12] and their fully corrected values obtained using equation [7.11], are shown in Figure 7.6(a) for a relatively long sample (13.8 mm in length). As expected from the strong temperature dependence involving heat loss, in accordance with the Stefan–Boltzmann law, the difference between the uncompensated values obtained from equation [7.9] and the compensated curves obtained based on equation [7.12] becomes more significant as the temperature increases, reaching approximately 31% at 300 K. Note that there is little difference between the fully corrected curve according to equation [7.11] and the compensated curve according to equation [7.12]. From the above results, the proposed compensation approach provides an accurate value of the Thomson coefficient in non-critical scenarios. Finally, it may be worth mentioning that the AC–DC method can be applied not only to a thin wire sample, but also to a thin-film sample fabricated on a substrate. It is likely that such a compensation mechanism also applies to the heat loss passing through a substrate from the thin-film sample to the heat reservoir, as long as the heat flows in only one direction (Fujiki 2021).

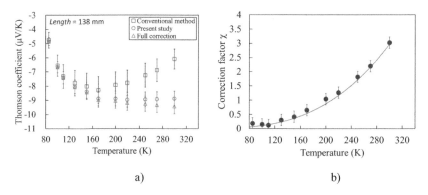

Figure 7.6. *(a) Measured Thomson coefficient of a long fine Pt wire as a function of temperature (l = 138 mm). (b) Measured and simulated correction term pyl/K, where an emissivity ε of 0.23 is used as the fitting parameter. Reprinted from Amagai et al. (2019b), with the permission of AIP Publishing. For a color version of this figure, see www.iste.co.uk/akinaga/thermoelectric1.zip*

Once the Thomson coefficient is measured, the Seebeck coefficient S is calculated from the measured Thomson coefficient using Kelvin's relation and a superconductor as a reference. Computation of the S value from the measured Thomson coefficient data involves the following steps: the value obtained from the superconductor reference ($YBa_2Cu_3O_{7-\delta}$) at 85 K is taken as the starting point, and subsequent calculations are performed by integrating the measured Thomson coefficient normalized with the temperature under the curve in Figure 7.6(a). The computed Seebeck coefficients with and without compensation are shown in Figure 7.7 along with the reference data (Moore and Graves 1973). The data computed from the compensated value show good agreement with the reference value within the expanded uncertainty.

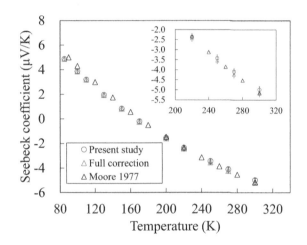

Figure 7.7. *Absolute Seebeck coefficient as a function of temperature, obtained using equation [7.12] and the fully corrected curve using equation [7.11], as well as the reference value for platinum (Moore and Graves 1973). The inset shows an enlarged view of the data between 200 K and 300 K. The error bar denotes the expanded uncertainty (k = 2), where k is the coverage factor. Reproduced from Amagai et al. (2019b), with the permission of AIP Publishing. For a color version of this figure, see www.iste.co.uk/akinaga/thermoelectric1.zip*

7.4. Summary and outlook

We have introduced the accurate measurement methods for the Thomson heat and the application to the precision measurement of the absolute Seebeck coefficient, also known as an absolute scale of thermoelectricity. From the recent developments of the Thomson coefficient measurements, the experiment for the measurement of the Thomson coefficient will be extremely simplified so that new techniques can be employed to provide local calibrations of a Seebeck voltage lead (measurement of the absolute Seebeck coefficient of the lead) to meet most of the laboratories' requirements. These advances will bring thermophysical property metrology much closer to the International System of Units (SI), and ensure the reliability of the characterization of thermoelectric materials, potentially including thermoelectric devices and modules. Moreover, the measurement uncertainty can undoubtedly be reduced. The presently accepted absolute scale of thermoelectricity was established more than 40 years ago. These new techniques may answer the question of whether there are errors in the present absolute scale. Combining the modern measurement standards and instruments, we believe that recent advances in techniques used to measure the Thomson heat represent useful tools to reevaluate the absolute scale of thermoelectricity.

7.5. References

Akinaga, H. (2020). Recent advances and future prospects in energy harvesting technologies. *Jpn J. Appl. Phys.*, 59, 110201.

Alleno, E., Bérardan, D., Byl, C., Candolfi, C., Daou, R., Decourt, R., Guilmeau, E., Hébert, S., Hejtmanek, J., Lenoir, B. et al. (2015). A round robin test of the uncertainty on the measurement of the thermoelectric dimensionless figure of merit of $Co_{0.97}Ni_{0.03}Sb_3$. *Rev. Sci. Instrum.*, 86, 011301.

Amagai, Y., Yamamoto, A., Akoshima, M., Fujiki, H., Kaneko, N.-H. (2015). AC/DC transfer technique for measuring Thomson coefficient: Toward thermoelectric metrology. *IEEE Trans. Instrum. Meas.*, 64(6), 1576–1581.

Amagai, Y., Shimazaki, T., Okawa, K., Fujiki, H., Kawae, T., Kaneko, N.-H. (2019a). Precise measurement of absolute Seebeck coefficient from Thomson effect using AC–DC technique. *AIP Advance.*, 9, 065312.

Amagai Y., Shimazaki T., Okawa K., Fujiki H., Kawae T., Kaneko N.-H. (2019b). High-accuracy compensation of radiative heat loss in Thomson coefficient measurement *Appl. Phys. Lett.*, 117(6), 063903.

Amagai, Y., Shimazaki, T., Okawa, K., Kawae, T., Fujiki, H., Kaneko, N.-H. (2020). Precise absolute Seebeck coefficient measurement and uncertainty analysis using high-T_c superconductors as a reference. *Rev. Sci. Instrum.*, 91, 014903.

Blatt, F.J. and Schroeder P.A. (eds) (1978). *Thermoelectricity in Metallic Conductors: International Conference on Thermoelectric Properties of Metallic Conductors, 1st, Michigan State University, 1977*. Springer, NewYork.

Borelius, G., Keesom, W.H., Johansson, C.H., Linde, J.O. (1932). Establishment of an absolute scale for the thermo-electric force. *Proc. Acad. Sci. Amst.*, 35, 10–14.

Borup, K.A., de Boor, J., Wang, H., Drymiotis, F., Gascoin, F., Shi, X., Chen, L., Fedorov, M.I., Müller, E., Iversen, B.B. et al. (2015). Measuring thermoelectric transport properties of materials. *Energy Environ. Sci.*, 8, 423–435.

Burkov, A.T., Heinrich, A., Konstantinov, P.P., Nakama, T., Yagasaki, K. (2001). Experimental set-up for thermopower and resistivity measurements at 100–1300 K. *Meas. Sci. Technol.*, 12, 264–272.

Callen, H.B. (1948). The application of Onsager's reciprocal relations to thermoelectric, thermomagnetic, and galvanomagnetic effects. *Phys. Rev.*, 73(11), 1349–1358.

Christian, J.W., Jan, J.-P., Pearson, W.B., Templeton, I.M. (1958). Thermo-electricity at low temperatures VI. A redetermination of the absolute scale of thermo-electric power of lead. *Proc. Roy. Soc. A*, 245(1241) 213–221.

Cusack, N. and Kendall, P. (1958). The absolute scale of thermoelectric power at high temperature. *Proc. Phys. Soc.*, 72, 898–901.

Duckworth, H.E. (1960). *Electricity and Magnetism*. Holt, Rinehart and Winston, New York.

Dunn, I.H., Daou, R., Atkinson, C. (2019). A straightforward 2ω technique for the measurement of the Thomson effect. *Rev. Sci. Instrum.*, 90, 024902.

Fujiki, H., Amagai, Y., Okawa, K., Harumoto, T., Kaneko, N.-H. (2021). Development on measurement method for Thomson coefficient of thin film. *Measurement: J. Int. Measur. Conf.*, 185.

Ginzburg, V.L. (1990). Thermoelectric effect in the superconducting states. *Sov. Phys. Us.*, 34(2) 101–107.

Lander, J.J. (1948). Measurements of Thomson coefficients for metals at high temperatures and of Peltier coefficients for solid–liquid interfaces of metals. *Phys. Rev.*, 74(4), 479–488.

Lenz, E., Edler, F., Ziolkowski, P. (2013). Traceable thermoelectric measurements of Seebeck coefficients in the temperature range from 300 K to 900 K. *Int. J. Thermophys.*, 34, 1975–1981.

Lowhorn, N.D., Wong-Ng, W., Lu, Z.Q., Thomas, E., Otani, M., Green, M., Dilley, N., Sharp, J., Tran, T.N. (2009). Development of a Seebeck coefficient Standard Reference Material. *Appl. Phys. A*, 96, 511–514.

MacDonald, D.K.C. (1962). *Thermoelectricity: An Introduction to the Principles*. John Wiley and Sons, Inc., New York.

Martin, J., Tritt, T.M., Uher, C. (2010). High temperature Seebeck coefficient metrology. *J. Appl. Phys.*, 108, 121101.

Martin, J., Lu, Z.-Q., Wong-Ng, W., Krylyuk, S., Wang, D., Ren, Z. (2021). Development of a high-temperature (295–900 K) Seebeck coefficient standard reference material. *J. Mater. Res.*, 36, 3339–3352.

Maxwell, G.M., Lloyd, J.N., Keller, D.V. (1967). Measurement of Thomson heat in metallic systems. *Rev. Sci. Instrum.*, 38(8), 1084–1089.

Miller, D.G. (1960). Thermodynamics of irreversible processes: The experimental verification of the onsager reciprocal relations. *Chem. Rev.*, 60(1), 15–37.

Moore, J.P. and Graves, R.S. (1973). Absolute Seebeck coefficient of platinum from 80 to 340 K and the thermal and electrical conductivities of lead from 80 to 400 K. *J. Appl. Phys.*, 44, 1174–1178.

Nettleton, H.R. (1912). On a method of measuriny the Thomson. *Proc. Phys. Soc. London*, 25, 44–65.

Nettleton, H.R. (1921). On a special apparatus for the measurement at various temperatures of the Thomson effect in wires. *Proc. Phys. Soc. London*, 34, 71–85.

Nolas, G.S., Sharp, J., Goldsmid, H.J. (2001). *Themoelectrics: Basic Principles and New Materials Developments*. Springer, New York.

Onserger, L. (1931a). Reciprocal relations in irreversible process I. *Phys. Rev.*, 37, 405–426.

Onserger, L. (1931b). Reciprocal relations in irreversible process II. *Phys. Rev.*, 38, 2265–2279.

Pearson, W.B. and Templeton, I.M. (1958). Thermo-electricity at low temperatures III. The absolute scale of thermo-electric power: A critical discussion of the present scale at low temperatures and preliminary measurements towards its redetermination. *Proc. Roy. Soc. A.*, 231(1187) 534–544.

Peltier, J.C. (1834). Nouvelles expériences sur la caloricité des courants éléctriques. *Ann. Chim. Phys.*, 56, 371–385.

Roberts, R.B. (1977a). Absolute scale of thermoelectricity. *Nature*, 265, 226–227.

Roberts, R.B. (1977b). The absolute scale of thermoelectricity I. *Philos. Mag.: J. Theor. Exp. Appl. Phys.*, 36(1), 91–107.

Roberts, R.B. (1981). The absolute scale of thermoelectricity II. *Philos. Mag. Part B*, 43(6), 1125–1135.

Roberts, R.B., Righini, F., Compton, R.C. (1985). The absolute scale of thermoelectricity III. *Philos. Mag. Part B*, 52(6), 1147–1163.

Rowe, D.M. (1995). *Thermoelectrics Handbook.* CRC Press, Boca Raton.

Seebeck, T.J. (1825). *Magnetische Polarisation der Metalle und Erze durch Temperatur.* Abhandlungen der physikalischen Klasse der Königlichen Akademie der Wissenschafften zu Berlin, Aus den Jahren 1822 und 1823. Extracts from four lectures delivered at the Academy of Sciences in Berlin on August 16, 1821, October 18, 25, 1821, and February 11, 1822.

Thomson, W. (1851). On a mechanical theory of thermo-electric currents, *Proc. R. Soc. Edinburgh*, 3, 91–98.

Thomson, W. (1854). On the dynamical theory of heat. Part V. Thermo-electric currents. *Trans. R. Soc. Edinburgh*, 21(1), 123–127.

Tritt, T.M. (2004). *Thermal Conductivity: Theory, Properties, and Applications.* Springer, New York.

Uchida, K., Murata, M., Miura, A., Iguchi, R. (2020). Observation of the magneto-Thomson effect. *Phys. Rev. Lett.*, 125, 106601.

Uher, C. (1987). Use of high-T_c superconductors for the determination of the absolute thermoelectric power. *J. Appl. Phys.*, 62, 4636–4638.

Wang, H., Porter, W.D., Böttner, H., König, J., Chen, L., Bai, S., Tritt, T.M., Mayolet, A., Senawiratne, J., Smith, C. et al. (2013). Transport properties of bulk thermoelectrics – An international round-robin study, Part I: Seebeck coefficient and electrical resistivity. *J. Electron. Mater.*, 42, 654–664.

Ziman, J.M. (1972). *Principles of the Theory of Solids*, 2nd edition. Cambridge University Press, Cambridge.

8

Thermal Diffusivity Measurement of Thin Films by Ultrafast Laser Flash Method

Tetsuya BABA[1], Takahiro BABA[2] and Takao MORI[1,2]
[1] *National Institute for Materials Science (NIMS), WPI-MANA, Japan*
[2] *Graduate School of Pure and Applied Sciences, University of Tsukuba, Japan*

8.1. Introduction

The flash method has been established as the standard method for measuring the thermal diffusivity of high density solid materials such as metals, alloys, ceramics and semiconductors (Parker et al. 1961; Righini et al. 1973; Baba et al. 2001; Akoshima et al. 2013). It is widely used and commercial equipment is available. The metrological standard for thermal diffusivity was established under the Metric convention by the Working Group 9 on Thermophysical Quantity of the International Bureau of Weights and Measures (*Bureau international des poids et mesures* – BIPM), Consultative Committee for Thermometry (CCT) (Baba 2010). Document standards for thermal diffusivity measurement have been published: ISO, ASTM International, etc. (Baba et al. 2014). Now, traceability of thermal diffusivity measurement for bulk materials has been globally established (Baba 2010, 2014).

For a color version of all the figures in this chapter, see www.iste.co.uk/akinaga/thermoelectric1.zip.

Thermoelectric Micro/Nano Generators 1,
coordinated by Hiroyuki AKINAGA, Atsuko KOSUGA,
Takao MORI and Gustavo ARDILA.
© ISTE Ltd 2023.

Due to the rapid increase in the importance of thin films in modern materials science and industry, there is a strong need to establish techniques for measuring the thermal properties of thin films (Cahill et al. 2003; Baba 2004; Dwyer et al. 2017; Petsagkourakis et al. 2018), and various experimental and theoretical studies have been conducted (Baba et al. 2011; Cahill et al. 2014; Yang et al. 2015; Volz et al. 2016).

In order to measure cross-plane heat transport properties of thin films with thickness from 10 nanometers to several micrometers, Paddock et al. invented the time-domain thermoreflectance method (TDTR method), which observes the surface temperature by the thermoreflectance method after impulse heating of the same surface by a femtosecond or picosecond pulse laser (Paddock et al. 1986). Most TDTR methods observe the cooling rate of the surface rather than the heat diffusion time over a certain length (Cahill et al. 2004; Collins et al. 2014). Therefore, instead of measuring the thermal diffusivity of the film, those methods basically measure the thermal effusivity of the second layer underneath the metal of the first layer, which is about 100 nanometers thick (Cahill et al. 2014).

As a different approach from the TDTR method, ultrafast laser flash methods with a rear side heating-surface temperature (RF) arrangement were developed to quantitatively measure the cross-plane thermal diffusivity of the thin film (Taketoshi et al. 1999, 2001; Baba et al. 2011). In the ultrafast laser flash method, transient change in the surface temperature of the thin film is observed, and the thermal diffusivity of the thin film is determined by adapting the mathematical model to the transient temperature changes (Baba 2009). The thin film is heated with an ultrashort pulse laser to observe transient temperature changes. Mode-Lock ultrashort pulse lasers oscillate periodic pulses with a fixed repetition frequency. Electrical delay technology allowed us to observe transient temperature changes for longer than the pulse interval (Taketoshi et al. 2005).

In order to analyze transient temperature change observed by the electrical delay technology, conventional models have to be modified to correctly represent the actual thermal thermoreflectance signal (Taketoshi et al. 2005; Baba et al. 2011). Conventional models assume that the thin film is heated with a single pulse, but in reality, it is heated with a periodic pulse. As a result, conventional models fit only a limited time range, not the full range of pulse intervals. To solve this problem, a linear correction approach that assumes that the signal base reduction is linear is commonly adopted (Taketoshi et al. 2005).

However, the actual base reduction is not linear, so this approximation is only valid for a limited signal. In this chapter, a new analysis method for fitting the thermoreflectance signal after the Fourier transform is presented (Baba et al. 2021). The periodic temperature response is expanded by the Fourier series to realize curve fitting of the thermoreflectance signal over the entire range of pulse intervals. Thermal diffusivity of thin films is determined reliably with small evaluated uncertainty by the ultrafast laser flash with the Fourier expansion analysis.

8.2. Laser flash method and ultrafast laser flash method

8.2.1. *Laser flash method*

Thermal conductivity λ is calculated by the formula $\lambda = \alpha c \rho$ from the specific heat capacity c and the density ρ after thermal diffusivity α was measured by the ultrafast laser flash method or the conventional laser flash method.

When thermal diffusivity of bulk material is measured by the laser flash method, as shown in Figure 8.1, the front face of the flat sample kept at a constant temperature is heated uniformly by a laser pulse (Parker et al. 1961). Heat diffuses one-dimensionally from the heated face to the opposite face, and eventually the temperature over the sample becomes uniform. Since the normalized temperature rise rate on the back surface of the sample is proportional to the thermal diffusivity and inversely proportional to the square of the thickness of the sample, thermal diffusivity is calculated from the thickness of the sample and the heat diffusion time.

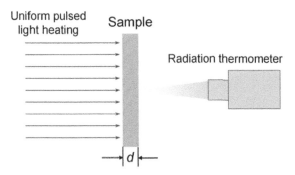

Figure 8.1. *Schematic of the laser flash method*

The following conditions are assumed as ideal (Parker et al. 1961; Baba et al. 2001):

1) duration of the laser pulse is negligibly short compared with the heat diffusion time;

2) the sample is adiabatic to the environment;

3) the sample's front face is heated uniformly;

4) the temperature change of the sample's rear face is measured precisely;

5) the sample is dense, uniform and opaque;

6) the change of thermal diffusivity due to the sample's temperature rise after the pulse heating is negligibly small.

Under the assumptions mentioned above, the temperature rise of the sample's rear face is expressed by the following equation:

$$T(t) = \Delta T \cdot \left[1 + 2\sum_{n=1}^{\infty}(-1)^n \exp\left(-(n\pi)^2 \frac{t}{\tau}\right)\right] = \frac{2}{b\sqrt{\pi t}}\sum_{n=0}^{\infty}\exp\left(-\frac{(2n+1)^2 \tau}{4t}\right) \quad [8.1]$$

where $\Delta T = Q/C$, Q is the total energy absorbed by the sample, C is the heat capacity of the sample, $b = \sqrt{\lambda c \rho}$ is the thermal effusivity of the sample and $\tau = d^2/\alpha$ is the heat diffusion time across the sample. The graph of equation [8.1] is shown in Figure 8.2 (Baba 2009).

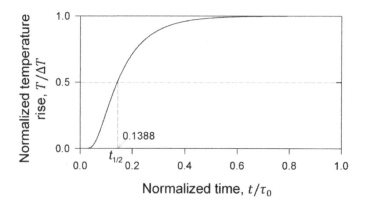

Figure 8.2. *Temperature rise of the rear face of the sample after impulse heating*

8.2.2. Ultrafast laser flash method

8.2.2.1. Picosecond pulsed light heating

Figure 8.3 shows the geometry of a pulsed light heating thermoreflectance apparatus under rear heating/front detection (RF) configuration. The temperature detection beam irradiates the front face of the thin film, and the heating beam irradiates the back face of the thin film through a transparent substrate.

The heat diffusion time of the entire thin film of a known thickness is measured, and the thermal diffusivity of the thin film is directly calculated from the heat diffusion time and the thickness of the thin film by adapting the RF configuration.

From the view point of derived property, the RF configuration is different from the front heating/front detection (FF) configuration, which observes temperature cooling after pulse heating, dominated by the thermal effusivity of the thin film and the substrate.

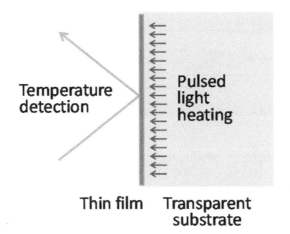

Figure 8.3. *Schematic of the pulsed light heating thermoreflectance measurement under rear heat/front detection (RF) configuration*

The thermal diffusivity of a thin film formed on a transparent substrate was measured by an ultrafast laser flash apparatus (PicoTR, NETZSCH-Gerätebau GmbH) in the research of this chapter.

The apparatus uses two fiber lasers: one for heating and one for temperature detection by thermoreflectance, as shown in Figure 8.4. The wavelength of the heating beam is 1,550 nm and the wavelength of the temperature detection beam is 775 nm, which is converted from the fundamental wavelength of 1,550 nm by the second harmonic generator (SHG). The differential photodiode detects the reflected light of the temperature detection beam from the surface of the sample. The diameter of the heating beam focused on the sample is 45 μm, and that of the temperature detection beam is 25 μm.

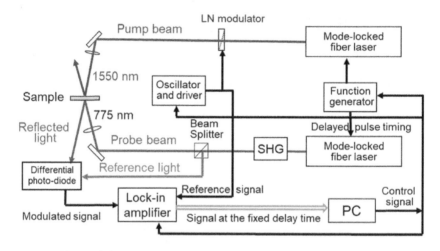

Figure 8.4. *Block diagram of an ultrafast laser flash apparatus*

The heated light is modulated by a lithium niobate modulator (LN modulator) at a frequency of 200 kHz and the signal of the reflected light is amplified at this frequency by the lock-in amplifier. Only the amplitude output of the lock-in amplifier was used to maintain the linearity of the signal with respect to temperature changes (Baba et al. 2021). The phase output of the lock-in amplifier was not used (Taketoshi et al. 2003).

The oscillating frequencies of the two fiber lasers are electrically controlled and synchronized at 20 MHz. Since the modulation frequency of 200 kHz is only 1% of the pulse repetition frequency of 20 MHz, the amplitude output of the lock-in amplifier is not distorted and linearity to the light intensity was kept.

The delay time of the temperature detection pulse train from the heating pulse train is controlled by the function generator. This electrical delay technique (Taketoshi et al. 2005) has enabled the observation of thermoreflectance signals over the entire interval (50 nanoseconds) between periodic pulses. However, the conventional "optical delay technology" is limited to observe thermoreflectance signals shorter than the repetition period of traditional mode-locked lasers from 12 to 13 ns (Paddock et al. 1986; Cahill et al. 2014; Collins et al. 2014).

The platinum thin film deposited on the fused quartz substrate was measured under RF configuration. The thin film is deposited by sputtering and its thickness is 100 nanometers. The black plot in Figure 8.5 is the thermoreflectance signal observed from the thin film. This signal is observed for the same thin film sample by a different measurement from the measurement in the reference (Baba et al. 2021). The pump laser has a nominal repeat rate of 20 MHz, so the periodic pulse interval is 50 nanoseconds. The sampling interval is 10 picoseconds.

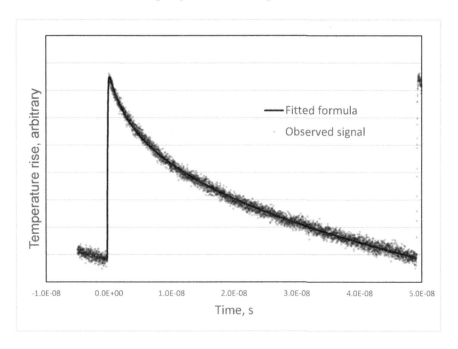

Figure 8.5. *RF thermoreflectance signal of platinum thin film (100 nm thick, fused quartz substrate) and regression curve in time domain (red line) by picosecond light pulse heating (Baba et al. 2021)*

8.2.2.2. Nanosecond pulsed light heating

In order to measure thicker films up to a few micrometers, nanosecond pulse lasers were used for pulsed light heating (Yagi et al. 2011; Kakefuda et al. 2017; Hinterleitner et al. 2019). The rear face of the thin film is heated by a laser beam through the transparent substrate. A CW semi-conductor laser is used for the probe beam and the reflected light proportional to the real time temperature change is detected by the high speed photodiode, as shown in Figure 8.6.

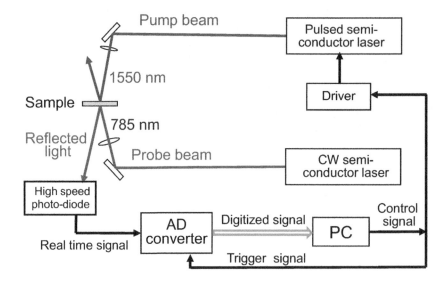

Figure 8.6. *Block diagram of nanosecond thermoreflectance measurement apparatus*

The pulse width of the pump pulse is 2 ns and the pulse interval is 20 µs. The thin film is formed on a disk-shaped transparent substrate with a diameter of 10 mm or a square-shaped transparent substrate with a side of 10 mm.

The black plot in Figure 8.7 shows the thermoreflectance signal for titanium nitride thin films which are 680 nm thick, supplied from National Metrology Institute of Japan (NMIJ) as a certified reference material (CRM) for cross-plane heat diffusion time of thin film by the ultrafast laser flash method of nanosecond pulsed light heating (Yagi et al. 2008).

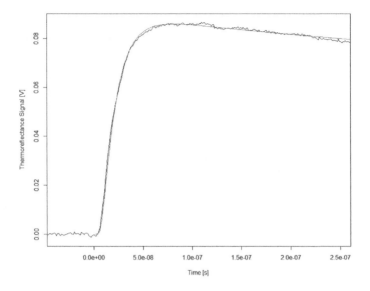

Figure 8.7. *RF thermoreflectance signal (black plot) of titanium nitride thin film (680 nm thick, fused quartz substrate) and regression curve in time domain (red curve) by picosecond light pulse heating (Kakefuda et al. 2017)*

8.3. Basic equation for data analysis

8.3.1. *Response function method*

The temperature response is expressed by the following Green's function (Baba 2009) after the surface of a semi-infinite solid material is impulse-heated by a unit intensity at time 0:

$$G(x,t\,|\,0,t') = \frac{1}{b\sqrt{\pi(t-t')}} \exp\left(-\frac{x^2}{4\alpha(t-t')}\right), \qquad [8.2]$$

where x is the distance from the surface to the observation position and t is the observed time. The thermophysical properties of solid materials are expressed as follows. α is the thermal diffusivity defined as $\alpha = \lambda/c\rho$. Thermal effusivity is defined as $b = \sqrt{\lambda c \rho} = c\rho\sqrt{\alpha}$: λ is the thermal conductivity, c is the specific heat capacity and ρ is the density.

When the front face of the flat layer is pulsewise heated, the transient temperature distribution inside the layer is given by the following Green's function (Baba 2009):

$$G(x,t\,|\,0,t') = \frac{1}{c\rho d}\left[1 + 2\sum_{n=1}^{v}\cos\frac{n\pi x}{d}\text{vexp}\left[-n^2\pi^2\frac{t-t'}{\tau}\right]\right]$$

$$= \frac{1}{b\sqrt{\pi(t-t')}}\sum_{n=-\infty}^{\infty}\exp\left(-\frac{(x-2nd)^2}{4\alpha(t-t')}\right)$$

[8.3]

The definition of $x, t, t', c, \rho, \alpha$ is the same as in equation [8.2]. d is the thickness of the layer and τ is the characteristic time of heat diffusion across the entire layer, defined as $\tau = d^2/\alpha$.

The temperature response of the position x at time t after being heated by any function $f(t)$ is expressed by the following convolution integral:

$$T(x,t) = \int_{-\infty}^{t} G(x,t\,|\,0,t')f(t')dt',$$

[8.4]

When the steady temperature T_0 is constant, the heat flow $q_f(t)$ flows onto the front face and $q_r(t)$ onto the rear face, the temperature response at the time t of each surface is expressed by the following equation (Baba 2009):

$$\mathbf{T}(t) = \mathbf{T}_0 + \int_0^t \mathbf{R}(t-t')\mathbf{q}(t')dt',$$

[8.5]

$$\mathbf{T}_0 = \begin{bmatrix} T_0 \\ T_0 \end{bmatrix},\ \mathbf{T}(t) = \begin{bmatrix} T_f(t) \\ T_r(t) \end{bmatrix},\ \mathbf{R}(t) = \begin{bmatrix} R_{ff}(t) & R_{fr}(t) \\ R_{fr}(t) & R_{rr}(t) \end{bmatrix},\ \mathbf{q}(t) = \begin{bmatrix} q_f(t) \\ q_r(t) \end{bmatrix},$$

where $\mathbf{R}(t)$ is the impulse response function matrix of the plate and the element $R_{ij}(t)$ is the temperature rise of the surface "i" at time "t" after heating the surface "j" by unit impulse at time 0. The subscript represents either the front or rear:

$$T_f(t) = T_0 + \int_0^t \left(R_{ff}(t-t')q_f(t') + R_{fr}(t-t')q_r(t') \right) dt', \qquad [8.6]$$

$$T_r(t) = T_0 + \int_0^t \left(R_{rf}(t-t')q_f(t') + R_{rr}(t-t')q_r(t') \right) dt'. \qquad [8.7]$$

These equations are simplified by considering the initial temperature T_0 as zero without losing generality.

The Laplace transform of $f(t)$ is expressed by $\tilde{f}(\xi)$ defined by the following equation, where ξ is a Laplace parameter:

$$\tilde{f}(\xi) = \int_0^\infty \exp(-\xi t) f(t) dt. \qquad [8.8]$$

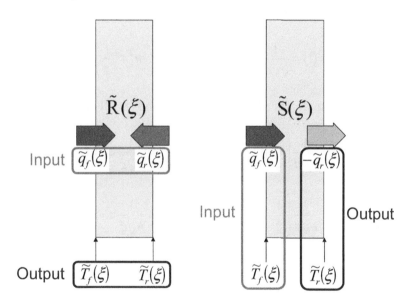

Figure 8.8. *Transfer function matrix (left) and quadruple matrix (right) (Baba et al. 2004)*

The Laplace transform of equation [8.5] is

$$\tilde{\mathbf{T}}(\xi) = \tilde{\mathbf{R}}(\xi) \cdot \tilde{\mathbf{q}}(\xi), \qquad [8.9]$$

$$\tilde{R}(\xi) = \begin{bmatrix} \tilde{R}_{ff}(\xi) & \tilde{R}_{fr}(\xi) \\ \tilde{R}_{rf}(\xi) & \tilde{R}_{rr}(\xi) \end{bmatrix}. \qquad [8.10]$$

where, $\tilde{R}(\xi)$ is the transfer matrix as shown on the left side of Figure 8.8.

8.3.2. *Uniform single layer*

Since Green's function of the uniform single-layer is expressed by equation [8.3], the elements of the impulse response matrix are expressed as follows (Baba 2009):

$$R_{ff}(t) = R_{rr}(t) = \frac{1}{c\rho d}\left[1 + 2\sum_{n=1}^{\infty}\exp\left(-n^2\pi^2 \frac{t}{\tau}\right)\right], \qquad [8.11]$$

$$R_{fr}(t) = R_{rf}(t) = \frac{1}{c\rho d}\left[1 + 2\sum_{n=1}^{\infty}(-1)^n \exp\left(-n^2\pi^2 \frac{t}{\tau}\right)\right]. \qquad [8.12]$$

The Laplace transforms of equations [8.11] and [8.12] are:

$$\tilde{R}_{ff}(\xi) = \tilde{R}_{rr}(\xi) = \frac{1}{c\rho d}\left[\frac{1}{\xi} + 2\sum_{n=1}^{\infty}\frac{1}{\xi + n^2\pi^2/\tau}\right] = \frac{1}{b\sqrt{\xi}}\coth\left(\sqrt{\xi\tau}\right) \qquad [8.13]$$

$$\tilde{R}_{fr}(\xi) = \tilde{R}_{rf}(\xi) = \frac{1}{c\rho d}\left[\frac{1}{\xi} + 2\sum_{n=1}^{\infty}\frac{(-1)^n}{\xi + n^2\pi^2/\tau}\right] = \frac{1}{b\sqrt{\xi}}\text{cosech}\left(\sqrt{\xi\tau}\right) \qquad [8.14]$$

8.3.3. *Quadruple matrix*

Equation [8.9] is expressed as follows:

$$\begin{bmatrix} \tilde{T}_f(\xi) \\ \tilde{T}_r(\xi) \end{bmatrix} = \begin{bmatrix} \tilde{R}_{ff}(\xi) & \tilde{R}_{fr}(\xi) \\ \tilde{R}_{rf}(\xi) & \tilde{R}_{rr}(\xi) \end{bmatrix}\begin{bmatrix} \tilde{q}_f(\xi) \\ \tilde{q}_r(\xi) \end{bmatrix} \qquad [8.15]$$

As shown in Figure 8.8, this equation is transformed by introducing a quadruple matrix that relates the heat flux density–temperature pair at the front-side with the same pair at the rear-side as follows:

$$\begin{bmatrix} -\tilde{q}_r(\xi) \\ \tilde{T}_r(\xi) \end{bmatrix} = \tilde{S}(\xi) \begin{bmatrix} \tilde{q}_f(\xi) \\ \tilde{T}_f(\xi) \end{bmatrix}, \qquad [8.16]$$

$$\tilde{S}(\xi) = \begin{bmatrix} \dfrac{\tilde{R}_{ff}(\xi)}{\tilde{R}_{fr}(\xi)} & -\dfrac{1}{\tilde{R}_{fr}(\xi)} \\ -\dfrac{\tilde{R}_{ff}(\xi)\cdot\tilde{R}_{rr}(\xi)-\tilde{R}_{fr}(\xi)\cdot\tilde{R}_{rf}(\xi)}{\tilde{R}_{fr}(\xi)} & \dfrac{\tilde{R}_{rr}(\xi)}{\tilde{R}_{fr}(\xi)} \end{bmatrix},$$

$$= \begin{bmatrix} \cosh\left(\sqrt{\tau\xi}\right) & -b\sqrt{\xi}\cdot\sinh\left(\sqrt{\tau\xi}\right) \\ -\dfrac{1}{b\sqrt{\xi}}\cdot\sinh\left(\sqrt{\tau\xi}\right) & \cosh\left(\sqrt{\tau\xi}\right) \end{bmatrix} \qquad [8.17]$$

where, the quadruple matrix $\tilde{S}(\xi)$ is a function of thermal effusivity b and heat diffusion time $\tau = d^2/\alpha$ (Baba 2009). It should be noted that $\tilde{q}_f(\xi)$ and $-\tilde{q}_r(\xi)$ are the vectors in the same direction as shown on the right side of Figure 8.8.

8.3.4. *Thin film/substrate model*

Figure 8.9 shows a thin film synthesized on a semi-infinite substrate. The parameters for specifying this thin film/substrate model are thickness of the thin film d_f, thermal diffusivity of the thin film α_f, thermal effusivity of the thin film b_f, thermal diffusivity of the substrate α_s and thermal effusivity of the substrate b_s (Baba 2009).

The temperature response at the film/substrate boundary, $T_S(t)$, is calculated as a convolution integral of the heat flow density flowing out to the substrate, $q_s(t)$, and the Green's function of the semi-infinite substrate shown in equation [8.2]:

$$T_s(t,0) = \int_0^t \frac{1}{b_s \sqrt{\pi(t-t')}} q_s(t')dt'. \qquad [8.18]$$

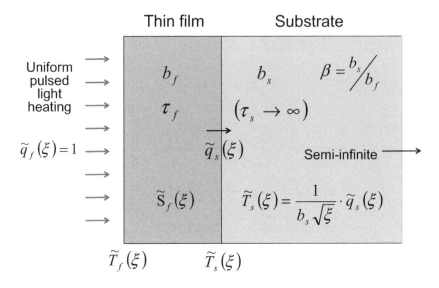

Figure 8.9. *Thin film/substrate temperature response, heat flow density and thermophysical properties (Baba 2009)*

The Laplace transform of this equation is

$$\tilde{T}_s(\xi) = \frac{1}{b_s \sqrt{\xi}} \cdot \tilde{q}_s(\xi) \qquad [8.19]$$

The pair of $q_s(t)$ and $T_s(t)$ is related to the pair of heat flow density, $q_f(t)$, and temperature, $T_f(t)$ to the surface of the thin film, via the quadruple matrix of the thin film $\tilde{S}_f(\xi)$, as shown in equation [8.16]:

$$\begin{bmatrix} \tilde{q}_s(\xi) \\ \tilde{T}_s(\xi) \end{bmatrix} = \tilde{S}_f(\xi) \cdot \begin{bmatrix} \tilde{q}_f(\xi) \\ \tilde{T}_f(\xi) \end{bmatrix} \qquad [8.20]$$

Substituting $q_f(t) = \delta(t)$, the delta function at time "0", the simultaneous equations [8.19] and [8.20] are solved, and $\tilde{T}_f(\xi)$ is expressed as follows:

$$\tilde{T}_f(\xi) = \frac{1}{b_s\sqrt{\xi}} \frac{\coth(\sqrt{\xi\tau_f}) + \beta}{\coth(\sqrt{\xi\tau_f}) + \beta^{-1}}, \qquad [8.21]$$

where the heat diffusion time of the entire film is defined as $\tau_f = d_f^2/\alpha_f$ and the thermal effusivity ratio of the substrate to the film is defined as $\beta = b_s/b_f$.

Substituting $\coth(x) = [(1+\exp(-2x))]/[1-\exp(-2x)]$, equation [8.21] is expressed as a function of $\exp(-2\sqrt{\xi\tau_f})$.

$$\begin{aligned}\tilde{T}_f(\xi) &= \frac{1}{b_f\sqrt{\xi}} \frac{1+\gamma\exp(-2\sqrt{\xi\tau_f})}{1-\gamma\exp(-2\sqrt{\xi\tau_f})} \\ &= \frac{1}{b_f\sqrt{\xi}}\left(1+2\sum_{n=1}^{\infty}\gamma^n \exp(-2n\sqrt{\xi\tau_f})\right)\end{aligned} \qquad [8.22]$$

where $\gamma = (b_f - b_s)/(b_f + b_s) = (1-\beta)/(1+\beta)$.

The surface temperature is obtained by the inverse Laplace transform of equation [8.22]:

$$T_f(t) = \frac{1}{b_f\sqrt{\pi t}}\left(1+2\sum_{n=1}^{\infty}\gamma^n \exp\left(-n^2\frac{\tau_f}{t}\right)\right) \qquad [8.23]$$

From equations [8.17], [8.20] and [8.22], the Laplace transform of the temperature response at the film/substrate boundary, $\tilde{T}_s(\xi)$, is calculated as follows:

$$\tilde{T}_s(\xi) = \frac{2}{(b_f + b_s)\sqrt{\xi}} \frac{\exp(-2\sqrt{\xi\tau_f})}{1-\gamma\exp(-2\sqrt{\xi\tau_f})},$$ [8.24]

$$= \frac{2}{(b_f + b_s)\sqrt{\xi}} \sum_{n=0}^{\infty} \gamma^n \exp\left(-(2n+1)\sqrt{\xi\tau_f}\right)$$

The temperature response at the film/substrate boundary, $T_s(t)$, is calculated by the inverse Laplace transform of equation [8.24]:

$$T_s(t) = \frac{2}{(b_f + b_s)\sqrt{\pi t}} \sum_{n=0}^{\infty} \gamma^n \exp\left(-(2n+1)^2 \frac{\tau_f}{t}\right)$$ [8.25]

According to the reciprocity of heat transfer, equation [8.25] under the front heating/rear detection (FR) configuration is equal to the temperature response under the rear heating/front detection (RF) configuration (Baba et al. 2020, 2021).

8.3.5. Temperature response after periodic pulse heating

8.3.5.1. Periodic pulse heating and Fourier series expansion

As mentioned in the introduction, the model function in the time domain assumes single pulse heating. However, the thin film surface is heated by periodic pulses in the pulsed light heating thermoreflectance method. When irradiated with a periodic laser pulse with a frequency of f_{rep}, the temperature response follows a periodic function along with the period of that frequency $\Delta T = 1/f_{rep}$. Since the periodic function can be expressed as a Fourier series, the periodic temperature response can also be expressed by the Fourier series.

When the mth discrete value of the thermoreflectance signal is y_m, the Fourier coefficient Y_n derived from the discrete Fourier Transform (DFT) is given by the following equation (Baba et al. 2021).

$$Y_n = \sum_{m=0}^{N-1} y_m \cdot \exp(-i2\pi v_n m \Delta t)$$ [8.26]

$$v_n = \frac{n}{N\Delta t} (n = 0,1,2,\dots, N-1)$$

where N is the number of samplings, Δt is the sampling interval and v_n is the frequency. Frequency v_n is defined as a multiple of the sampling rate $1/\Delta t$.

However, its inverse discrete Fourier transform (IDFT) is given by the following equation:

$$y_m = \frac{1}{N}\sum_{n=0}^{N-1} Y_n \cdot \exp(i2\pi v_n m\Delta t) \qquad [8.27]$$

According to the sampling theorem, frequency components that exceed the Nyquist rate $1/2\Delta t$ are alias.

Therefore, equation [8.27] can be converted as follows:

$$y_m = \frac{1}{N}[Y_0 + \sum_{n=1}^{(N-1)/2} Y_n \cdot \exp(i2\pi v_n m\Delta t) + \sum_{n=1}^{(N-1)/2} \overline{Y_n} \cdot \exp(i2\pi v_{N-n} m\Delta t)] \qquad [8.28]$$

where $\overline{Y_n}$ is a conjugate of Y_n.

Equation [8.28] can be thought of as a Fourier series of a periodic function with a period of $N\Delta t$. The period $N\Delta t$ corresponds to the interval of periodic pulses ΔT.

We already know that the thermoreflectance signal can be expressed by a transfer function $\tilde{Y}(\xi)$, which is $\tilde{T}_f(\xi)$ of equation [8.22] under the front heating/fear detection (FF) configuration and $\tilde{T}_s(\xi)$ of equation [8.24] under the front heating/rear detection (FR) configuration. Therefore, the Fourier coefficient Y_n follows the transfer function $\tilde{Y}(\xi)$ in the frequency domain.

If we define the complex frequency as $\xi_n = i2\pi v_n$, the regression model can be expressed as follows:

$$Y_n = \tilde{Y}(v_n, \hat{k}', \hat{\tau}_f, \hat{\gamma}) + \varepsilon \qquad [8.29]$$

where $\hat{k}', \hat{\tau}_f$ and $\hat{\gamma}$ are the estimates of the constant of proportionality k', heat diffusion time τ_f and ratio of virtual heat source γ, respectively. ε is an error term that expresses the residual. This regression model is valid in the frequency domain below the Nyquist rate. The coefficient at frequency zero Y_0 is excluded. Note that the four parameters $v_n, \hat{k}', \hat{\tau}_f, \hat{\gamma}$ are real numbers, but $Y_n, \tilde{Y}(v_n, \hat{k}', \hat{\tau}_f, \hat{\gamma})$ and ε are complex numbers.

8.3.5.2. *Fourier coefficient and Laplace transform of analytical formula*

Figure 8.10 shows the relationship between the temperature response of single pulse heating T_{single} and periodic pulse heating $T_{periodic}$. $T_{periodic}$ is understood as an accumulation of T_{single}, where we consider that T_{single} and $T_{periodic}$ are the continuous function of time t instead of the discrete value.

$$T_{periodic}(t) = \sum_{m=0}^{\infty} T_{single}(t + m\Delta T) \qquad [8.30]$$

Considering only the range of pulse intervals, the Fourier coefficient of order n is expressed by:

$$\hat{T}_{periodic}(\nu_n) = \int_0^{\Delta T} T_{periodic}(t) \exp(-i2\pi\nu_n t) dt$$

$$= \sum_{m=0}^{\infty} \left[\int_0^{\Delta T} T_{single}(t + m\Delta T) \exp\left(-i2\pi n \frac{t}{\Delta T}\right) dt \right]$$

$$= \sum_{m=0}^{\infty} \left[\int_0^{\Delta T} T_{single}(t + m\Delta T) \exp\left(-i2\pi n \frac{t+m\Delta T}{\Delta T}\right) dt \right] \qquad [8.31]$$

It should be noted that $\exp(-i2\pi nm)$ is equal to 1.

If the variable is changed from t to $t' = t + m\Delta T$, the equation is expressed as follows:

$$\hat{T}_{periodic}(\nu_n) = \sum_{m=0}^{\infty} \left[\int_{m\Delta T}^{(m+1)\Delta T} T_{single}(t') \exp\left(-i2\pi n \frac{t'}{\Delta T}\right) dt' \right]$$

$$= \int_0^{\infty} T_{single}(t') \exp(-i2\pi\nu_n t') dt' \qquad [8.32]$$

This means that the Fourier coefficient of the temperature response of periodic pulse heating corresponds to the Fourier transform of the temperature response of single pulse heating. If we redefine the complex variable as $\xi_n = i2\pi\nu_n$, $\hat{T}_{periodic}(\nu_n)$ can be expressed by $T_{single}(t)$ as follows:

$$\hat{T}_{periodic}(\nu_n) = \int_0^{\infty} T_{single}(t') \exp(-\xi_n t') dt' \qquad [8.33]$$

It should be noted that ξ_n is a complex number, whereas v_n is a real number.

Finally, $\hat{T}_{periodic}(v_n)$ is expressed by the Laplace transform of temperature response of single pulse heating $\tilde{T}_{single}(\xi)$ as follows:

$$\hat{T}_{periodic}(v_n) = \tilde{T}_{single}(\xi_n) \tag{8.34}$$

$T_{single}(t)$ corresponds to the function, $T_f(t)$ of [8.23], or $T_s(t)$ of [8.25].

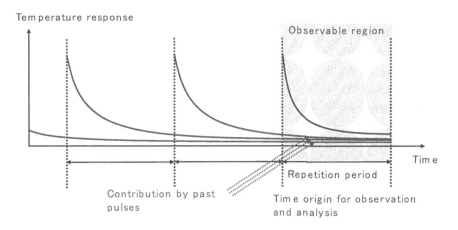

Figure 8.10. *Temperature response of a single pulse heating and periodic pulse heating (Baba et al. 2021)*

After determining the model function by regression analysis in the frequency domain, the thermal diffusivity and the thermal effusivity of the thin film, and the thermal effusivity and thermal effusivity of the substrate can be determined as follows:

$$\alpha_f = \frac{d_f^2}{\hat{\tau}_f}, b_f = \sqrt{\alpha_f c_f \rho_f}, b_s = b_f \frac{1-\hat{\gamma}}{1+\hat{\gamma}} \tag{8.35}$$

where d_f is the film thickness, c_f is the specific heat capacity and ρ_f is the density of the thin film. Note that an estimate of proportionality constant k' is not important and only serves as a fitting parameter.

When we consider the front face heating/rear detection (FR) configuration, which is equivalent to the rear face heating/front (RF) detection configuration, the periodic response in the time domain can be calculated as a Fourier series using the Fourier coefficient $\widehat{Y_n}$ with the determined parameters substituted as follows:

$$\widehat{Y_n} = \tilde{Y}(v_n, \hat{k}', \hat{\tau}_f, \hat{\gamma}) = \frac{\hat{k}'}{\sqrt{i2\pi v_n}} \cdot \frac{\exp(-\sqrt{i2\pi v_n \hat{\tau}_f})}{1-\hat{\gamma}\exp(-2\sqrt{i2\pi v_n \hat{\tau}_f})} \quad [8.36]$$

$$\hat{y}(t) = \frac{1}{N}[Y_0 + \sum_{n=1}^{(N-1)/2} \widehat{Y_n} \cdot \exp(i2\pi v_n t) + \sum_{n=1}^{(N-1)/2} \overline{\widehat{Y_n}} \cdot \exp(i2\pi v_{N-n} t)] \quad [8.37]$$

Note that the square root $\sqrt{i2\pi v_n}$ and $\sqrt{i2\pi v_n \hat{\tau}_f}$ in equation [8.36] are multivalued because they contain the imaginary unit i. The square root of a complex number is always calculated as $(1+i)/\sqrt{2}$ to keep the branch cuts of complex functions consistent.

The temperature response can be expressed as a discrete expression considering the sampling interval as follows (Baba et al. 2021):

$$\widehat{y_m} = \frac{1}{N}[Y_0 + \sum_{n=1}^{(N-1)/2} \widehat{Y_n} \cdot \exp(i2\pi v_n m\Delta t) + \sum_{n=1}^{(N-1)/2} \overline{\widehat{Y_n}} \times \exp(i2\pi v_{N-n} m\Delta t)] \quad [8.38]$$

The coefficients at frequency zero contain only information about the baseline. The corresponding values already derived from the DFT can be used.

Figures 8.11(a) and (b) show the theoretical temperature response curve under the RF configuration after periodic pulse heating calculated from equation [8.37]. The time scale is normalized by the periodic pulse intervals, and the temperature scale is arbitrarily normalized. Figure 8.11(a) shows the temperature response curves when the heat diffusion times of the entire thin film are different with $\gamma = 1$. When the thermal effusivity of the substrate is 0, it corresponds to the heat insulation boundary condition of the thin film. The heat diffusion time τ_f is normalized by ΔT, and a dimensionless parameter $\Phi = \tau_f/\Delta T$ is introduced. Figure 8.11(b) shows a different temperature response curve for different γ with $\Phi = 0.1$.

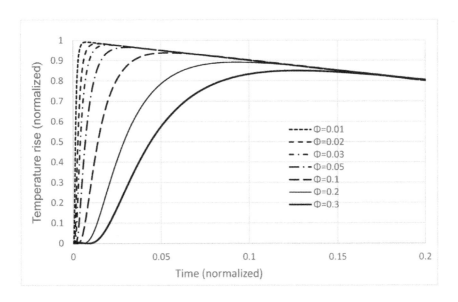

Figure 8.11(a). *Theoretical temperature responses depend on Φ (γ = 1) (Baba et al. 2021)*

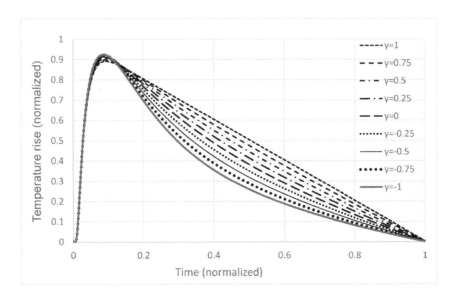

Figure 8.11(b). *Theoretical temperature responses depend on γ (Φ = 0.1) (Baba et al. 2021)*

8.4. Analysis of observed temperature response

8.4.1. *Picosecond pulsed light heating*

The Fourier coefficient Y_n was calculated by DFT from the thermoreflectance signal in the range of 0 seconds to 50 nanoseconds. Y_n is a complex number consisting of a real part and an imaginary part. Complex numbers can also be represented by absolute values and arguments.

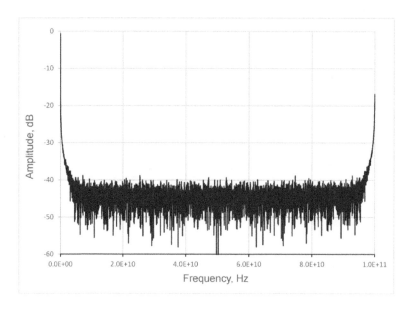

Figure 8.12. *Magnitude of frequency components of thermoreflectance signal in Figure 8.5 (Baba et al. 2021)*

Figure 8.12 shows the absolute value of the Fourier coefficient $|Y_n|$. This plot can be thought of as a frequency response. Note that the Fourier coefficient shows symmetry with respect to the Nyquist rate 50 GHz.

Since this sample is modeled as a semi-infinitely thick film on a substrate, as shown in Figure 8.9, the Laplace transform of the temperature response after single pulse heating is expressed by equation [8.24]. A curve fitting program using the nonlinear least squares method was used. The regression analysis was applied only to the absolute value of the Fourier coefficient $|Y_n|$ as follows:

$$|Y_n| = |\tilde{Y}(v_n, \hat{k}', \hat{\tau}_f, \hat{\gamma})| + \varepsilon' \qquad [8.39]$$

Figure 8.13 shows a regression curve in the frequency domain. As mentioned above, the curve fitting was done in the range less than the Nyquist rate. This time, the high- frequency component of 5 GHz or more is omitted. As shown in Figure 8.13, by fitting equation [8.29] to the absolute value of the Fourier coefficient, it was determined to be $\hat{\tau}_f$ at 6.47×10^{-10} [s] and $\hat{\gamma}$ at 0.676. The red curve in Figure 8.5 is the theoretical curve in the time domain derived from equation [8.36]. Note that by introducing three fitting parameters, the signal and theoretical curve observed over the iteration period match each other.

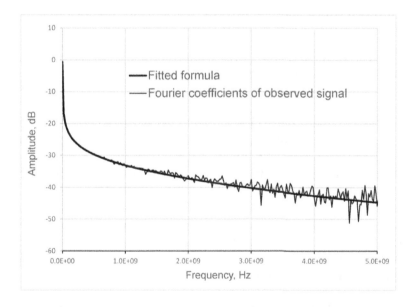

Figure 8.13. *Regression curve in frequency domain (Baba et al. 2021)*

Since the thickness of the platinum layer is 100 nanometers, the thermal diffusivity of the platinum thin film was calculated at 1.55×10^{-5} m^{2}s^{-1} by equation [8.35]. Taking the literature value of the specific heat capacity of platinum and the density as $c_f = 133$ [J/(kg · K)], the density as

$\rho_f = 21{,}500$ [kg/m^3] and the thermal effusivity of platinum as 11,240 [J/(s$^{0.5}$m^{2}K)], the fuzed quartz substrate b_s was calculated as 2,175 [J/(s$^{0.5}$m^{2}K)].

8.4.2. Nanosecond pulsed light heating

Since the observed temperature response for the TiN thin film of 680 nm thick measured by nanosecond pulse heating returns the original level in a much shorter time than the pulse duration period, an analytical equation after single pulse heating is applicable. Therefore, the black plot in Figure 8.7 was fitted by equation [8.25] and shown by the red curve in the figure.

The observed heat diffusion time is 1.347×10^{-7} s, which agreed to the reference value of CRM "1.397×10^{-7} s" within 4%. The thermal diffusivity of the titanium nitride thin film was calculated as 3.4×10^{-6} m^{2}s^{-1}.

8.5. Metrological standard and traceability for measurements of thin film thermophysical properties

The pulsed light heating thermoreflectance method of front face heating/rear face detection (FR) configuration and rear face heating/front face detection (RF) configuration is an absolute measurement method under ideal measurement conditions. After systematic and comprehensive investigation, the National Institute of Advanced Industrial Science and Technology, National Metrology Institute of Japan (NMIJ) established a heat diffusion time standard for thin films (Yagi et al. 2008). All measuring instruments, such as the sampling period and electrical sensitivity of transient memory are traceable to the metrological standard. Measurement results are critically evaluated based on Guide to the Expression of Uncertainty in Measurement (GUM).

The heat diffusion time standard is provided by NMIJ, with a relative uncertainty of 8% for picosecond pulse heating and 5% for nanosecond pulse heating, expressed in expanded measurement uncertainty of coverage factor 2. Based on these measurement standards, NMIJ provides thermal diffusivity reference material, which meets the requirements of a certified reference material quality management system (Yagi et al. 2008; Baba 2010; Baba et al. 2010).

As shown in Figure 8.14, the reliability of such thermophysical property data extends the concept of uncertainty evaluation and traceability for "measurement" to standard values for standard materials and standard data, and then to general thermophysical property data. It can be guaranteed by expanding to uncertainty assessment and guarantee of traceability (Baba et al. 2014).

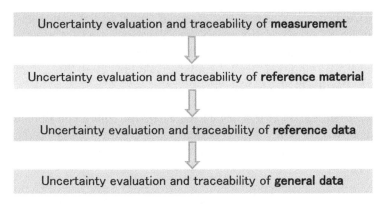

Figure 8.14. *Generalization of uncertainty evaluation and traceability to from measurement to data (Baba et al. 2014)*

8.6. Application of measurement from industrial to basic physics

The ultrafast laser flash method can measure a wide variety of thin films. If the optical properties of the thin film do not meet the requirements of the method, coating with a metallic thin film such as molybdenum is one solution (Baba 2009), then, semiconductors, carbon materials, oxide thin films, nitrides, carbides and polymers.

Thin films developed for thermoelectric applications have been measured by the ultrafast laser flash method under both RF configuration and FF configuration (Kakefuda et al. 2017; Hinterleitner et al. 2019; Bourgès et al. 2020; Gao et al. 2020; Latronico et al. 2021a, 2021b; Lima et al. 2021; Liu et al. 2021; Wang et al. 2021; Lei et al. 2022; Peng et al. 2022; Van Hoang et al. 2022). Transparent conductive films for flat panel displays (Yagi et al. 2005; Ashida et al. 2007, 2009; Oka et al. 2010), phase change materials for optical media and P-RAM (Kuwahara et al. 2006, 2007a, 2007b, 2007c, 2009), and polymer thin films for organic electroluminescence (Oka et al.

2010b) have also been measured for example, showing the usefulness of this method.

The ultrafast laser flash method has been applied to thermal diffusivity measurement under high pressure in a diamond anvil cell (Yagi et al. 2010), and also to thin films synthesized on a silicon substrate (Firoz et al. 2010).

A focused beam system, for example, with a few microns spatial resolution in the in-plane dimension, has also been used to measure small crystals, or potentially site selective in-plane thermal conductivity measurements of inhomogeneous samples (Kakefuda et al. 2017; Piotrowski et al. 2018; Pawula et al. 2019).

For the first time, heat transfer across a 100-nm thin metal thin film was observed down to the liquid helium temperature region, where the temperature response deviates from the heat diffusion equation (Nakamura et al. 2011).

A highly reliable thermal design can be realized by simulation using the thermal diffusivity value of these highly reliable thin films and the boundary thermal resistance between the thin films measured by the ultrafast laser flash method.

As described in this chapter, the ultrafast laser flash method has been developed to be a physically and mathematically rigorous method to measure and determine the thermal diffusivity of thin films. This is vital information for a wide variety of applications, as introduced above. The ultrafast laser flash method is also evolving and becoming more versatile by the day, and further advancements are eagerly awaited.

8.7. References

Akoshima, M. and Baba, T. (2006). Study on a thermal-diffusivity standard for laser flash method measurements. *International Journal of Thermophysics*, 27(4), 1189–1203. doi: 10.1007/s10765-006-0091-9.

Akoshima, M., Hay, B., Zhang, J., Chapman, L., Baba, T. (2013). International comparison on thermal-diffusivity measurements for iron and isotropic graphite using the laser flash method in CCT-WG9. *International Journal of Thermophysics*, 34(5), 763–777. doi: 10.1007/s10765-012-1209-x.

Ashida, T., Miyamura, A., Sato, Y., Yagi, T., Taketoshi, N., Baba, T., Shigesato, Y. (2007). Effect of electrical properties on thermal diffusivity of amorphous indium zinc oxide films. *Journal of Vacuum Science & Technology A*, 25(4), 1178–1183. doi: 10.1116/1.2743644.

Ashida, T., Miyamura, A., Oka, N., Sato, Y., Yagi, T., Taketoshi, N., Baba, T., Shigesato, Y. (2009). Thermal transport properties of polycrystalline tin-doped indium oxide films. *Journal of Applied Physics*, 105(7), 073709. doi: 10.1063/1.3093684.

Baba, T. (2004). General needs on nanoscale thermal metrology and the Japanese program on this subject. *Proceedings of Therminic Workshop 2004*, Sophia Antipolis, Côte d'Azur.

Baba, T. (2009). Analysis of one-dimensional heat diffusion after light pulse heating by the response function method. *Japanese Journal of Applied Physics*, 48(5), 05EB04. doi: 10.1143/jjap.48.05eb04.

Baba, T. (2010). Measurements and data of thermophysical properties traceable to a metrological standard. *Metrologia*, 47(2), S143.

Baba, T. and Akoshima, A. (2014). A social system for production and utilization of thermophysical quantity data. *Synthesiology – English Edition*, 7(2), 49–64.

Baba, T. and Ono, A. (2001). Improvement of the laser flash method to reduce uncertainty in thermal diffusivity measurements. *Measurement Science and Technology*, 12(12), 2046–2057. doi: 10.1088/0957-0233/12/12/304.

Baba, T., Yamada, N., Taketoshi, N., Watanabe, H., Akoshima, M., Yagi, T., Abe, H., Yamashita, Y., Baba, T. (2010). Research and development of metrological standards for thermophysical properties of solids in the National Metrology Institute of Japan. *High Temperatures-High Pressures*, 39(4), 279–306.

Baba, T., Taketoshi, N., Yagi, T. (2011). Development of ultrafast laser flash methods for measuring thermophysical properties of thin films and boundary thermal resistances. *Japanese Journal of Applied Physics*, 50(11), 11RA01. doi: 10.1143/jjap.50.11ra01.

Baba, T., Baba, T., Mori, T. (2020). Experimental investigation of reciprocity of temperature response across two layer samples by flash method. *Review of Scientific Instruments*, 91(1), 014905. doi: 10.1063/1.5124799.

Baba, T., Baba, T., Ishikawa, K., Mori, T. (2021). Determination of thermal diffusivity of thin films by applying Fourier expansion analysis to thermo-reflectance signal after periodic pulse heating. *Journal of Applied Physics*, 130(22), 225107. doi: 10.1063/5.0069375.

Bourgès, C., Sato, N., Baba, T., Baba, T., Ohkubo, I., Tsujii, N., Mori, T. (2020). Drastic power factor improvement by Te doping of rare earth-free CoSb3-skutterudite thin films. [10.1039/D0RA02699A]. *RSC Advances*, 10(36), 21129–21135. doi: 10.1039/d0ra02699a.

Cahill, D.G. (2004). Analysis of heat flow in layered structures for time-domain thermoreflectance. *Review of Scientific Instruments*, 75(12), 5119–5122. doi: 10.1063/1.1819431.

Cahill, D.G., Goodson, K., Majumdar, A. (2001). Thermometry and thermal transport in micro/nanoscale solid-state devices and structures. *Journal of Heat Transfer*, 124(2), 223–241. doi: 10.1115/1.1454111.

Cahill, D.G., Ford, W.K., Goodson, K.E., Mahan, G.D., Majumdar, A., Maris, H.J., Mrlin, R., Phillpot, S.R. (2003). Nanoscale thermal transport. *Journal of Applied Physics*, 93(2), 793–818. doi: 10.1063/1.1524305.

Cahill, D.G., Braun, P.V., Chen, G., Clarke, D.R., Fan, S., Goodson, K.E., Keblinski, P., King, W.P., Mahan, G.D., Majumdar, A. et al. (2014). Nanoscale thermal transport II: 2003–2012. *Applied Physics Reviews*, 1(1), 011305. doi: 10.1063/1.4832615.

Cezairliyan, A., Baba, T., Taylor, R. (1994). A high-temperature laser-pulse thermal diffusivity apparatus. *International Journal of Thermophysics*, 15(2), 317–341. doi: 10.1007/bf01441589.

Collins, K.C., Maznev, A.A., Cuffe, J., Nelson, K.A., Chen, G. (2014). Examining thermal transport through a frequency-domain representation of time-domain thermoreflectance data. *Review of Scientific Instruments*, 85(12), 124903. doi: 10.1063/1.4903463.

Firoz, S.H., Yagi, T., Taketoshi, N., Ishikawa, K., Baba, T. (2010). Direct observation of thermal energy transfer across the thin metal film on silicon substrates by a rear heating–front detection thermoreflectance technique. *Measurement Science and Technology*, 22(2), 024012. doi: 10.1088/0957-0233/22/2/024012.

Gao, W., Liu, Z., Baba, T., Guo, Q., Tang, D.-M., Kawamoto, N., Bauer, E., Tsujii, N., Mori, T. (2020). Significant off-stoichiometry effect leading to the N-type conduction and ferromagnetic properties in titanium doped Fe_2VAl thin films. *Acta Materialia*, 200, 848–856. doi: 10.1016/j.actamat.2020.09.067.

Hinterleitner, B., Knapp, I., Poneder, M., Shi, Y., Müller, H., Eguchi, G., Eisenmenger-Sittner, C., Stöger-Pollach, M., Kakefuda, Y., Kawamoto, N. et al. (2019). Thermoelectric performance of a metastable thin-film Heusler alloy. *Nature*, 576(7785), 85–90. doi: 10.1038/s41586-019-1751-9.

Kakefuda, Y., Yubuta, K., Shishido, T., Yoshikawa, A., Okada, S., Ogino, H., Kawamoto, N., Baba, T., Mori, T. (2017). Thermal conductivity of PrRh4.8B2, a layered boride compound. *APL Materials*, 5(12), 126103. doi: 10.1063/1.5005869.

Kobayashi, K. and Baba, T. (2009). Extension of the response time method and the areal heat diffusion time method for one-dimensional heat diffusion after impulse heating: Generalization considering heat sources inside of multilayer and general boundary conditions. *Japanese Journal of Applied Physics*, 48(5), 05EB05. doi: 10.1143/jjap.48.05eb05.

Kuwahara, M., Suzuki, O., Taketoshi, N., Yamakawa, Y., Yagi, T., Fons, P., Tsutumi, K., Suzuki, M., Fukaya, T., Tominaga, J. et al. (2006). Measurements of temperature dependence of optical and thermal properties of optical disk materials. *Japanese Journal of Applied Physics*, 45(2B), 1419–1421. doi: 10.1143/jjap.45.1419.

Kuwahara, M., Suzuki, O., Taketoshi, N., Yagi, T., Fons, P., Tominaga, J., Baba, T. (2007a). Thermal conductivity measurements of Sb–Te alloy thin films using a nanosecond thermoreflectance measurement system. *Japanese Journal of Applied Physics*, 46(10A), 6863–6864. doi: 10.1143/jjap.46.6863.

Kuwahara, M., Suzuki, O., Yamakawa, Y., Taketoshi, N., Yagi, T., Fons, P., Tominaga, J., Baba, T. (2007b). Temperature dependence of the thermal properties of optical memory materials. *Japanese Journal of Applied Physics*, 46(6B), 3909–3911. doi: 10.1143/jjap.46.3909.

Kuwahara, M., Suzuki, O., Yamakawa, Y., Taketoshi, N., Yagi, T., Fons, P., Fukaya, T., Tominaga, J., Baba, T. (2007c). Measurement of the thermal conductivity of nanometer scale thin films by thermoreflectance phenomenon. *Microelectronic Engineering*, 84(5), 1792–1796. doi: 10.1016/j.mee.2007.01.178.

Kuwahara, M., Suzuki, O., Tsutsumi, K., Yagi, T., Taketoshi, N., Kato, H., Simpson, R.E., Suzuki, M., Tominaga, J., Baba, T. (2009). Measurement of refractive index, specific heat capacity, and thermal conductivity for Ag6.0In4.5Sb60.8Te28.7 at high temperature. *Japanese Journal of Applied Physics*, 48(5), 05EC02. doi: 10.1143/jjap.48.05ec02.

Latronico, G., Mele, P., Artini, C., Manfrinetti, P., Pan, S.W., Kawamura, Y., Sekine, C., Singh, S., Takeuchi, T., Baba, T. et al. (2021a). Investigation on the power factor of skutterudite Smy(FexNi1−x)4Sb12 thin films: Effects of deposition and annealing temperature. *Materials*, 14(19), 5773.

Latronico, G., Singh, S., Mele, P., Darwish, A., Sarkisov, S., Pan, S.W., Kawamura, Y., Sekine, C., Baba, T., Mori, T. et al. (2021b). Synthesis and characterization of Al- and SnO2-doped ZnO thermoelectric thin films. *Materials*, 14(22), 6929.

Lei, Y., Qi, R., Chen, M., Chen, H., Xing, C., Sui, F., Gu, L., He, W., Zhang, Y., Baba, T. et al. (2022). Microstructurally tailored thin β-Ag2Se films toward commercial flexible thermoelectrics. *Advanced Materials*, 34(7), 2104786. doi: 10.1002/adma.202104786.

Lima, M.S.L., Aizawa, T., Ohkubo, I., Baba, T., Sakurai, T., Mori, T. (2021). High power factor in epitaxial Mg2Sn thin films via Ga doping. *Applied Physics Letters*, 119(25), 254101. doi: 10.1063/5.0074707.

Liu, S., Li, G., Lan, M., Baba, T., Baba, T., Mori, T., Wang, Q. (2021). Thermoelectric performance enhancement of film by pulse electric field and multi-nanocomposite strategy. *Small*, 17(40), 2100554. doi: 10.1002/smll.202100554.

Ma, W., Miao, T., Zhang, X., Kohno, M., Takata, Y. (2015). Comprehensive study of thermal transport and coherent acoustic-phonon wave propagation in thin metal film–substrate by applying picosecond laser pump–probe method. *The Journal of Physical Chemistry C*, 119(9), 5152–5159. doi: 10.1021/jp512735k.

Nakamura, F., Taketoshi, N., Yagi, T., Baba, T. (2010). Observation of thermal transfer across a Pt thin film at a low temperature using a femtosecond light pulse thermoreflectance method. *Measurement Science and Technology*, 22(2), 024013. doi: 10.1088/0957-0233/22/2/024013.

O'Dwyer, C., Chen, R., He, J.-H., Lee, J., Razeeb, K.M. (2017). Scientific and technical challenges in thermal transport and thermoelectric materials and devices. *ECS Journal of Solid State Science and Technology*, 6(3), N3058.

Oka, N., Arisawa, R., Miyamura, A., Sato, Y., Yagi, T., Taketoshi, N., Baba, T., Shigesato, Y. (2010a). Thermophysical properties of aluminum oxide and molybdenum layered films. *Thin Solid Films*, 518(11), 3119–3121. doi: 10.1016/j.tsf.2009.09.180.

Oka, N., Kato, K., Yagi, T., Taketoshi, N., Baba, T., Ito, N., Shigesato, Y. (2010b). Thermal diffusivities of Tris(8-hydroxyquinoline)aluminum and N,N'-Di(1-naphthyl)-N,N'-diphenylbenzidine thin films with sub-hundred nanometer thicknesses. *Japanese Journal of Applied Physics*, 49(12), 121602. doi: 10.1143/jjap.49.121602.

Paddock, C.A. and Eesley, G.L. (1986). Transient thermoreflectance from thin metal films. *Journal of Applied Physics*, 60(1), 285–290. doi: 10.1063/1.337642.

Parker, W.J., Jenkins, R.J., Butler, C.P., Abbott, G.L. (1961). Flash method of determining thermal diffusivity, heat capacity, and thermal conductivity. *Journal of Applied Physics*, 32(9), 1679–1684. doi: 10.1063/1.1728417.

Pawula, F., Daou, R., Hébert, S., Lebedev, O., Maignan, A., Subedi, A., Kakefuda, Y., Kawamoto, N., Baba, T., Mori, T. (2019). Anisotropic thermal transport in magnetic intercalates Fe_xTiS_2. *Physical Review B*, 99(8), 085422. doi: 10.1103/PhysRevB.99.085422.

Peng, Y., Miao, L., Liu, C., Song, H., Kurosawa, M., Nakatsuka, O., Back, S., Rhyee, J.-S., Murata, M., Tanemura, S. et al. (2022). Constructed ge quantum dots and sn precipitate sigesn hybrid film with high thermoelectric performance at low temperature region. *Advanced Energy Materials*, 12(2), 2103191. doi: 10.1002/aenm.202103191.

Petsagkourakis, I., Tybrandt, K., Crispin, X., Ohkubo, I., Satoh, N., Mori, T. (2018). Thermoelectric materials and applications for energy harvesting power generation. *Science and Technology of Advanced Materials*, 19(1), 836–862.

Piotrowski, M., Franco, M., Sousa, V., Rodrigues, J., Deepak, F.L., Kakefuda, Y., Kawamoto, N., Baba, T., Owens-Baird, B., Alpuim, P. et al. (2018). Probing of thermal transport in 50 nm thick PbTe nanocrystal films by time-domain thermoreflectance. *The Journal of Physical Chemistry C*, 122(48), 27127–27134. doi: 10.1021/acs.jpcc.8b04104.

Righini, F. and Cezairliyan A. (1973). Pulse method of thermal diffusivity measurements. A review. *High Temperatures-High Pressures*, 5, 481.

Taketoshi, N., Baba, T., Ono, A. (1999). Observation of heat diffusion across submicrometer metal thin films using a picosecond thermoreflectance technique. *Japanese Journal of Applied Physics*, 38(11A), L1268–L1271. doi: 10.1143/jjap.38.l1268.

Taketoshi, N., Baba, T., Ono, A. (2001). Development of a thermal diffusivity measurement system for metal thin films using a picosecond thermoreflectance technique. *Measurement Science and Technology*, 12(12), 2064–2073. doi: 10.1088/0957-0233/12/12/306.

Taketoshi, N., Baba, T., Schaub, E., Ono, A. (2003). Homodyne detection technique using spontaneously generated reference signal in picosecond thermoreflectance measurements. *Review of Scientific Instruments*, 74(12), 5226–5230. doi: 10.1063/1.1628840.

Taketoshi, N., Baba, T., Ono, A. (2005). Electrical delay technique in the picosecond thermoreflectance method for thermophysical property measurements of thin films. *Review of Scientific Instruments*, 76(9), 094903. doi: 10.1063/1.2038628.

Taketoshi, N., Yagi, T., Baba, T. (2009). Effect of synthesis condition on thermal diffusivity of molybdenum thin films observed by a picosecond light pulse thermoreflectance method. *Japanese Journal of Applied Physics*, 48(5), 05EC01. doi: 10.1143/jjap.48.05ec01.

Van Hoang, D., Tuan Thanh Pham, A., Baba, T., Huu Nguyen, T., Bao Nguyen Le, T., Dieu Thi Ung, T., Hong, J., Bae, J-S., Park, H., Park, S. et al. (2022). New record high thermoelectric ZT of delafossite-based CuCrO2 thin films obtained by simultaneously reducing electrical resistivity and thermal conductivity via heavy doping with controlled residual stress. *Applied Surface Science*, 583, 152526. doi: 10.1016/j.apsusc.2022.152526.

Volz, S., Shiomi, J., Nomura, M., Miyazaki, K. (2016). Heat conduction in nanostructured materials. *Journal of Thermal Science and Technology*, 11(1), JTST0001–JTST0001. doi: 10.1299/jtst.2016jtst0001.

Wang, Y., Pang, H., Guo, Q., Tsujii, N., Baba, T., Baba, T., Mori, T. (2021). Flexible n-type abundant chalcopyrite/PEDOT:PSS/graphene hybrid film for thermoelectric device utilizing low-grade heat. *ACS Applied Materials & Interfaces*, 13(43), 51245–51254. doi: 10.1021/acsami.1c15232.

Yagi, T., Tamano, K., Sato, Y., Taketoshi, N., Baba, T., Shigesato, Y. (2005). Analysis on thermal properties of tin doped indium oxide films by picosecond thermoreflectance measurement. *Journal of Vacuum Science & Technology A*, 23(4), 1180–1186. doi: 10.1116/1.1872014.

Yagi, T., Ohta, K., Kobayashi, K., Taketoshi, N., Hirose, K., Baba, T. (2010). Thermal diffusivity measurement in a diamond anvil cell using a light pulse thermoreflectance technique. *Measurement Science and Technology*, 22(2), 024011. doi: 10.1088/0957-0233/22/2/024011.

Yamashita, Y., Yagi, T., Baba, T. (2011). Development of network database system for thermophysical property data of thin films. *Japanese Journal of Applied Physics*, 50(11), 11RH03. doi: 10.1143/jjap.50.11rh03.

Yang, F. and Dames, C. (2015). Heating-frequency-dependent thermal conductivity: An analytical solution from diffusive to ballistic regime and its relevance to phonon scattering measurements. *Physical Review B*, 91(16), 165311. doi: 10.1103/PhysRevB.91.165311.

List of Authors

Hiroyuki AKINAGA
National Institute of Advanced
Industrial Science and
Technology (AIST)
Device Technology Research
Institute
Tsukuba
Japan

Yasutaka AMAGAI
National Metrology Institute of Japan
National Institute of Advanced
Industrial Science and Technology
(AIST)
Japan

Gustavo ARDILA
Grenoble Alpes University
and
IMEP – LaHC
France

Takahiro BABA
Graduate School of Pure and
Applied Sciences
University of Tsukuba
Japan

Tetsuya BABA
National Institute for Materials
Science (NIMS)
WPI-MANA
Japan

Tsuyohiko FUJIGAYA
Department of Applied Chemistry
Graduate School of Engineering
and
The World Premier International
Research Center Initiative
International Institute of Carbon
Neutral Energy Research
and
Department of Chemical Engineering
Graduate School of Engineering
and
Center for Molecular Systems
Kyushu University
Japan

Prashun GORAI
Colorado School of Mines
Golden, CO
USA

Kazuki IMASATO
Global Zero Emission Research Center
National Institute of Advanced Industrial Science and Technology (AIST)
Japan

Priyanka JOOD
Global Zero Emission Research Center
National Institute of Advanced Industrial Science and Technology (AIST)
Japan

Holger KLEINKE
Department of Chemistry and Waterloo Institute for Nanotechnology
University of Waterloo
Canada

Atsuko KOSUGA
Osaka Metropolitan University
Department of Physical Science
Graduate School of Science
Japan

Takao MORI
National Institute for Materials Science (NIMS)
WPI-MANA
and
Graduate School of Pure and Applied Sciences
University of Tsukuba
Japan

Yoshiyuki NONOGUCHI
Faculty of Materials Science and Engineering
Kyoto Institute of Technology
Japan

Michihiro OHTA
Global Zero Emission Research Center
National Institute of Advanced Industrial Science and Technology (AIST)
Japan

Ichiro TERASAKI
Department of Physics
Nagoya University
Japan

Michael TORIYAMA
Northwestern University
Evanston, IL
USA

Index

A, B

AC–DC method, 185, 190, 192, 194
American Society for Testing and Materials (ASTM), 201
anisotropy, 162
band
   engineering, 9
   structure, 9
BiCuSeO, 132, 135

C, D

Ca_2RuO_4, 139
carbon nanotubes (CNTs), 150, 151, 153, 154, 157–159, 161–176
certified reference material (CRM), 224
chalcopyrite, 106, 107
chemical doping, 154, 156, 175
Chevrel-phase, 94, 105
colusite, 94, 103, 104, 107
computational, 17–19, 24, 27, 29, 35–37, 39, 40, 45, 46, 51, 52
conjugated network, 151
contact layer, 107, 108
convolution integral, 210
copper, 74, 75, 77–80, 84
data, 29, 44, 45, 49, 50
dataset, 48
defect calculations, 41, 43
density, 203
   functional theory, 31, 45
dimensionless parameter, 220
Discrete Fourier Transform (DFT), 216, 217

E, F

electrical delay, 202
enhancement, 5, 6, 9–11
fiber laser, 206
figure of merit, 73, 75, 76, 82, 84
flash method, 201–203, 205, 208, 225, 226
   laser, 201
   ultrafast laser, 202
flexible, 150, 151, 163, 169–171, 173, 175
Fourier (*see also* Discrete Fourier Transform)
   coefficient, 217
   expansion, 203
   series, 203
   transform, 203
   front heating/rear detection (FR) configuration, 216

G, H

Green's function, 209
Guide to the expression of Uncertainty in Measurement (GUM), 224
heat diffusion
  equation, 226
  time, 203
heat transfer
  equation, 192
  reciprocity of, 216
Heikes formula, 132, 133, 139
high-
  spin state, 130, 138
  throughput, 17, 18, 28, 29, 31, 34, 43, 45, 46, 52
homologous series, 100

I, K

impulse response, 210, 212
  function, 210
International Bureau of Weights and Measures (*Bureau international des poids et mesures* – BIPM), 201
International Organization for Standardization (ISO), 201
inverse design, 42
Kelvin formula, 128, 129, 134, 139

L, M

Laplace transform, 211, 212, 214–216, 218, 219, 222
  inverse, 215
layered
  cobalt oxides, 132–135, 137
  rhodium oxides, 137, 142
  sulfides, 94, 97, 98
low
  -dimensional, 151
  -spin state, 130, 137
machine learning, 18, 44, 45, 51

magnon drag, 9, 11
metrology (*see also* thermophysical property metrology)
  standard unit, 201
misfit-layered sulfide, 97, 98
modeling, 17, 18, 24, 26, 32, 40, 41, 51, 52
Mott gap, 126, 128

N, O, P

Na_xCoO_2, 131, 133, 137
nanostructuring, 5–7
organic materials, 162
patterning, 167, 168, 172–175
Peltier effect, 184
periodic pulse heating, 216, 218–220
perovskite-type structure, 130, 131
phase transition, 75, 76, 78
phonon scattering, 6, 8
polymers, 150–156, 162, 169, 170, 172, 174
power factor, 4, 8–11
precision measurement, 195

Q, R, S

quadruple matrix, 213
rare-earth sulfide, 94
rear heating/front detection (RF) configuration, 205
sample's
  front face, 204
  rear face, 204
Seebeck
  absolute coefficient, 183–189, 195, 196
  coefficient, 4, 8–11
  effect, 184, 186, 189
selenide, 74, 75, 81–84
silver, 75, 77–79, 84
single pulse heating, 218
specific heat capacity, 203
structure-sorted, 154

sulfide, 74, 75, 77, 78, 81–84
superconductor, 186, 195
superionic conductor, 94, 101, 108, 109

T, U

telluride, 74, 80–83
tetrahedrite, 94, 103, 104
theory, 17–20, 24, 26, 28, 29, 31, 36, 39, 40, 45
thermal
   conductivity, 4–8, 74–84, 203, 209, 226
   diffusivity, 201–205, 209, 213, 219, 223, 224, 226
   effusivity, 202, 204, 205, 209, 213, 215, 219, 220, 224
   ratio, 215
thermoelectric, 3, 4, 11

thermoelectricity
   absolute scale of, 184, 185, 187–189, 195, 196
thermophysical property metrology, 184, 196
thermoreflectance, 202, 203, 205–209, 216, 217, 222, 224
   method, 202, 216, 224
   signal, 202, 203, 207, 208, 216, 217, 222
thin film/substrate model, 213
Thomson effect, 183, 184, 186, 189–193
time-domain thermoreflectance (TDTR) method, 202
traceability, 225
transfer function matrix, 211, 212
transport calculations, 26–29
uncertainty, 203

Summary of Volume 2

Preface
Hiroyuki AKINAGA, Atsuko KOSUGA and Takao MORI

Introduction
Hiroyuki AKINAGA, Atsuko KOSUGA and Takao MORI

Part 1. Material Challenges and Novel Effects

Chapter 1. Reliability and Durability of Thermoelectric Materials and Devices: Present Status and Strategies for Improvement
Congcong XU, Hongjing SHANG, Zhongxin LIANG, Fazhu DING and Zhifeng REN

    1.1. Introduction
    1.2. Thermoelectric material stability
    1.3. $Mg_3(Sb, Bi)_2$
    1.4. Zn_4Sb_3
    1.5. Skutterudites
    1.6. $Cu_{2-x}X$ (X = S, Se, Te)
    1.7. GeTe
    1.8. Outlook on thermoelectric materials stability
    1.9. Thermoelectric device design analysis
        1.9.1. Thermal stress analysis
        1.9.2. Interface analysis, design and fabrication
    1.10. Advanced thermoelectric module case studies
        1.10.1. Bi_2Te_3
        1.10.2. $Mg_3(Sb, Bi)_2$
        1.10.3. GeTe
        1.10.4. Skutterudites
    1.11. Summary and outlook
    1.12. References

Chapter 2. Effect of Microstructure in Understanding the Electronic Properties of Complex Materials
Chenguang FU, Chaoliang HU, Qi ZHANG, Airan LI and Tiejun ZHU

2.1. Introduction
2.2. Basic principles of electronic transport parameters
    2.2.1. Solid solutions
    2.2.2. Intrinsic defects
    2.2.3. Grain boundary
    2.2.4. Texture
2.3. Summary
2.4. References

Chapter 3. Thermoelectric Nanowires
Olga CABALLERO-CALERO and Marisol MARTÍN-GONZÁLEZ

3.1. Introduction
3.2. Nanowires: a way to enhance thermoelectric efficiency
3.3. Fabrication of thermoelectric nanowires
3.4. Measurement of thermoelectric properties in nanowires
3.5. Nanowire-based thermoelectric devices
3.6. Interconnected 3D nanowire networks
3.7. Summary and outlook
3.8. References

Chapter 4. Impact of Chemical Doping or Magnetism in Model Thermoelectric Sulfides
Sylvie HÉBERT, Ramzy DAOU and Antoine MAIGNAN

4.1. Introduction
4.2. TiS_2: intercalation chemistry to combine power factor optimization and lattice thermal conductivity degradation
4.3. Magnetism and thermoelectricity in sulfides
4.4. Conclusion
4.5. References

Chapter 5. Thermoelectric Generation Using the Anomalous Nernst Effect
Akito SAKAI and Satoru NAKATSUJI

    5.1. Thermoelectric conversion – Seebeck effect and anomalous Nernst effect (ANE)
    5.2. Physics of topological magnets
        5.2.1. Transverse electrical and thermal conductivity driven by Berry curvature
        5.2.2. Magnetic Weyl semimetals, Weyl magnets

    5.2.3. Type-II Weyl semimetals
    5.2.4. Nodal line magnets
5.3. Experimental realization of the giant anomalous Nernst effect
    5.3.1. Weyl antiferromagnets Mn_3X (X = Sn, Ge)
    5.3.2. Weyl ferromagnet Co_2MnGa
    5.3.3. Nodal-web ferromagnets Fe_3X (X = Ga, Al)
5.4. Summary and prospects
5.5. Acknowledgment
5.6. References

Chapter 6. A Comprehensive Review of Phonon Engineering
Bin XU, Harsh CHANDRA, and Junichiro SHIOMI

6.1. Introduction
    6.1.1. Thermal conductivity
    6.1.2. Phonons in thermal transport
6.2. Methodology of phonon engineering
    6.2.1. Computational method for thermal conduction and phonon properties
    6.2.2. Experimental method for nano-/micro-scale heat conduction characterization
    6.2.3. Direct measurement of phonon properties through phonon scattering
    6.2.4. Phonon engineering for low thermal conductivity
    6.2.5. Intrinsic low thermal conductivity in complex lattice structure
    6.2.6. Low thermal conductivity by nanostructures
    6.2.7. Coherent phonon engineering in superlattice
6.3. Summary and future prospects
6.4. References

Part 2. Toward Device Applications

Chapter 7. The Current State of Thermoelectric Technologies and Applications with Prospects
Slavko BERNIK

7.1. Introduction
7.2. Thermoelectric materials
7.3. Thermoelectric devices – structure, materials, fabrication technology
7.4. Summary
7.5. References

Chapter 8. Processing of Thermoelectric Transition Metal Silicides Towards Module Development
Sylvain LE TONQUESSE, Mathieu PASTUREL, Franck GASCOIN and David BERTHEBAUD

8.1. Introduction
8.2. Recent progress on the process of thermoelectric transition metals silicide
    8.2.1. Synthesis of mesostructured silicides through magnesiothermic reduction
    8.2.2. Synthesis of higher manganese silicide through wet ball milling
    8.2.3. Issues of MnSi striations and thermal stability on thermoelectric performance of doped higher manganese silicide
    8.2.4. Upscaling processes, the examples of additive manufacturing and RGS process
8.3. Towards contacts and device developments
8.4. References

Chapter 9. Application of the Thermoelectrics; Past, Present and Future
Hirokuni HACHIUMA

9.1. Introduction
9.2. Thermoelectric module
9.3. TEC application for refrigerator and cooler
9.4. TEC for electronic components
    9.4.1. TEC for optical communication
    9.4.2. Multi-stage TEC for optical sensors
9.5. TEC for semiconductor manufacturing
9.6. TEG application
    9.6.1. TEG for energy harvesting (EH)
    9.6.2. TEG for stand-alone power source
    9.6.3. TEG for waste heat recovery
9.7. Conclusion
9.8. References

Printed and bound by CPI Group (UK) Ltd, Croydon, CR0 4YY
19/12/2023

08211922-0001